SpringerBriefs in Environmental Science

SpringerBriefs in Environmental Science present concise summaries of cutting-edge research and practical applications across a wide spectrum of environmental fields, with fast turnaround time to publication. Featuring compact volumes of 50 to 125 pages, the series covers a range of content from professional to academic. Monographs of new material are considered for the SpringerBriefs in Environmental Science series.

Typical topics might include: a timely report of state-of-the-art analytical techniques, a bridge between new research results, as published in journal articles and a contextual literature review, a snapshot of a hot or emerging topic, an in-depth case study or technical example, a presentation of core concepts that students must understand in order to make independent contributions, best practices or protocols to be followed, a series of short case studies/debates highlighting a specific angle.

SpringerBriefs in Environmental Science allow authors to present their ideas and readers to absorb them with minimal time investment. Both solicited and unsolicited manuscripts are considered for publication.

More information about this series at http://www.springer.com/series/8868

Soumyananda Dinda

Climate Friendly Goods and Technologies in Asia

Opportunities for Trade

 Springer

Soumyananda Dinda
Department of Economics
University of Burdwan
Burdwan, West Bengal, India

ISSN 2191-5547   ISSN 2191-5555   (electronic)
SpringerBriefs in Environmental Science
ISBN 978-3-030-02474-1   ISBN 978-3-030-02475-8   (eBook)
https://doi.org/10.1007/978-3-030-02475-8

Library of Congress Control Number: 2018958910

This Springer imprint is published by the registered company Springer Nature Switzerland AG
The registered company address is: Gewerbestrasse 11, 6330 Cham, Switzerland

Preface

Emissions of greenhouse gases associated with continued economic expansion need to be controlled by adopting sustainable production and consumption that emphasizes energy-efficient technologies. Countries produce and export climate-friendly goods and technologies (CFGT), which have relatively less adverse impacts on the environment. Liberalized trade can make available such goods and clean technologies for countries that have no access to such CFGT, or wherein domestic industries are unable to produce them at sufficient scales or at affordable prices. Through trade, countries in Asia provide affordable renewable technologies and make them widely available for mitigating climate change. This book explicitly focuses on potential trade opportunities of climate-friendly goods and technologies for countries in Asia and South Asia, provides trade performance of climate-friendly goods in the early twenty-first century in Asia and estimates potential trade gaps in Asia. It has several chapters with major focus on CFGT trade performance, regional orientation, trade analysis and estimation of its potential trade gaps in the abovesaid regions, and covers major research areas of climate-friendly goods and technology aspects and their classified product groups and subgroups. Trade volume is examined by identifying and tracking the unique HS code associated with each technology or product under the Harmonized Commodity Description and Coding System. This book also helps to clarify India's position on global warming and regional efforts to mitigate climate change in the international level.

This book has four parts. Part 1 provides conceptual ideas, related literature and data descriptions and methodologies. Part 2 shows the trade performance measuring several indicators. Part 3 depicts regional orientation, and Part 4 analyses and estimates the potential trade for climate-friendly goods and technologies in Asia, South Asia and India, mapping the movement of CFGT.

I am grateful to Mia Mikic for her suggestion and conceptualization on climate-friendly goods and technological trade. I would like to thank the seminar participants at Indian Institute of Technology, Madras; Chandragupt Institute of Management, Patna; Indian Statistical Institute, Madras School of Economics;

Delhi School of Economics; Institute of Economic Growth; Jadavpur University; Oldenburg University; Burdwan University; Sidho-Kanho-Birsha University; Maastricht School of Management; Birmingham University; Nottingham University; and the University of Bath where parts of this book were read out.

Burdwan, India   Soumyananda Dinda

Acknowledgements

I appreciate the support and efforts of all who were involved in making this book possible, which includes the referees, the editor and editorial team of Springer Publication. I gratefully acknowledge the assistance provided by the editorial staff for this book.

I must heartily thank my friends, Martine Wermelinger, Arijit Mukherji and Tirthankar Banerji, for their constant encouragement.

It is needless to acknowledge the most needed support and encouragement given to me by my family members, especially my wife and my eldest brother.

Contents

Abbreviation

APTA	Asia Pacific Trade Agreement
ASEAN	Association of South East Asian Nation
CCT	Clean Coal Technology
CFGT	Climate Friendly Goods and Technology
DC	Developed Country
EEL	Energy Efficient Lighting
EGS	Environmental Goods and Services
GDP	Gross Domestic Product
GHG	Green House Gas
LDC	Less Developed Country
OECD	Organisation for Economic Cooperation and Development
RCA	Reveal Comparative Advantage
SAARC	South Asian Association for Regional Cooperation
SPVS	Solar Photovoltaic System
USD	US Dollar
WE	Wind Energy
WTO	World Trade Organisation

Part I
Introducing Climate Friendly Goods and Technology

Chapter 1
Introduction

Abstract This chapter introduces the concept of climate friendly goods and technology (CFGT) and technology transfer through trade channel. It briefly discusses the export-led development paradigm shifting with trade diversity in Asia and threats of climate change in the twenty-first century. It also introduces the shifting economic crisis (Asia crisis to global crisis) associated with the global climate policy regimes. It raises issues of climate constraint and estimates potential trade opportunity with study plan.

Keywords Climate change · Export-led development · Cost of development · Environmental degradation · Capacity and capability · GHG · EGS · CFGT · Potential opportunity · Doha round · Efficient technology · Clean product · Energy-efficient technology · Asian financial crisis · Global financial crisis · Southeast Asia · Emerging market · East Asia · Renewable energy

1.1 Introduction

Asian newly industrializing economies demonstrate export-led growth since the 1970s. However, export-led development has gained momentum in newly global order in the early twenty-first century. Emerging economies like India and China have managed to achieve high economic growth rate with significant reduction of poverty which is based on *export-led development paradigm*. Trade would remain important in sustaining fast economic growth and development of Asia and more specific to South Asia and Southeast Asia. The experiences of East Asian economies reconfirm that export is the most important source of economic growth. In this context, it should be mentioned that export of industrial products positively contributes to economic growth, while it has certain negative impact on environment. Export promotion of manufacturing products contributes negatively to overall environment; in other words, it degrades environment. Truly, cost of development is observed and measured in terms of environmental degradation.

3

S. Dinda, *Climate Friendly Goods and Technologies in Asia*, SpringerBriefs in Environmental Science, https://doi.org/10.1007/978-3-030-02475-8_1

Now, the global concern is to maintain a balance between export and environment for both developed and developing countries (Mani 2014). Particularly, emissions of greenhouse gases associated with continued economic expansion need to be put under control by adopting sustainable production and consumption. In order to avoid conflicts between trade and environmental degradation, developed nations should assist developing countries in terms of improving their capacities and capabilities such that they can avail the advantage of new opportunities which are emerging in the world trade (World Bank 2008, Mani 2014). Trade in environmental goods is one such new opportunity (Nguyen and Kalirajan 2015), which has emerged from failures of the Doha round trade negotiation meetings. The Doha round did not explicitly cover categories of 'environmental monitoring and assessment equipment' and 'cleaner or more resource-efficient technologies and products'. This study focuses on the abovesaid categories and highlights on the early twenty-first-century trade status of such items and its potential opportunities in Asia.

Truly, the composition of the trade pattern may shift from high energy-intensive to less energy-intensive and/or clean products over time to fulfil the changing world demand due to growing environmentalists' agitation. In this context, Asian countries need to increase energy efficiency which is possible through switching energy sources from fossil fuel to renewable energy. Energy-efficient technologies are required for this purpose. Asia emerges as a dynamic business leader in the world, and the centre of gravity has shifted to Asia. Performance of Asian economy was excellent with trade diversity during 2002–2008.

Asian Financial Crisis to Global Financial Crisis Truly, Asia emerges a laboratory for experiment of economic activities in the neo-liberalism during 1997–2008. Southeast Asia is the most important laboratory which generated the Asian financial crisis 1997–1998 and examined the interventions to insulate from the Asian crisis 1997–1998 and collapsed export to developed countries during the global financial crisis 2008–2009. Southeast Asian economies are originator of Asian financial crisis 1997–1998; however, they were victims of the global financial crisis 2008–2009, which originated in the USA. Interestingly, the growth rates of Southeast Asian economies rebounded back by 2000 and 2010, respectively. The objective of this study is to investigate trade performance and opportunities between periods of *Asian financial crisis 1997–1998 and global financial crisis 2008–2009*. The global financial crisis 2008–2009 has strongly demonstrated the economic fortunes of Asia, the USA and the rest of the world. The crisis was transmitted to industrial and emerging market economies through both financial and trade channels. Declining demand for imports among advanced economies transmitted the crisis to export-reliant countries in Asia. Major trade dynamics with product diversity was observed between abovesaid two crises, i.e. during 1997–2008.

East Asia and Southeast Asia regions have taken a lead role in development and export of energy-efficient technologies. Asian countries provide affordable renewable technologies through trade and make them available widely for mitigating global climate change issues. So, there is possible emerging business opportunity to improve energy efficiency by adopting renewable energy sources and technologies. This study provides an overview on threats of climate change issues, trade channel for climate mitigation strategy and potential trade opportunities for climate friendly goods and technologies (CFGT) in Asia and its subregions during 1997–2008. However, major focus has been given on the global financial crisis in 2008–2009.

1.1.1 Climate Change and Threats to Human Civilization

Climate change refers to any significant change in the climate over time. It is a significant shift of climate lasting for an extended period of time. In the natural process, the climate has always been changing slowly. The current impact of human activities is causing the climate to change in an unnatural way and at a faster pace than ever before. Climate change is a global phenomenon which has certain impacts on the world. The human activity-induced climate change is causing shifts in the normal climatic conditions such as rainfall, temperature, etc., which in turn have impact on natural environment and living beings.

Greenhouse effect
Normally, the sunlight is the warm energy which is released from the Sun and travels through the Earth's atmosphere and hits the surface of the Earth. Part of it is reflected, and part of this warm energy is released back into the space and partly bounces back into the atmosphere. Carbon dioxide (CO_2) and other (methane, nitrous oxide, etc.) gases in the atmosphere trap warm energy (or heat) of the sunlight. The heat-trapping gases are termed as *greenhouse gases* (GHGs). The process of GHG trapping the Sun's heat is called the *greenhouse effect*. GHGs are essential to make the Earth warm enough for existence and/or survival of lives. Natural GHGs provide us comfortable environment with life support systems on the Earth.

Since industrialization, human activities have been releasing more and more heat-trapping gases in the atmosphere. Over time, human activities have increased the concentration of GHG in our atmosphere, trapping more heat which in turn is the main cause of increasing temperature. This rising temperature is the cause of the global warming. The *Intergovernmental Panel on Climate Change (IPCC)* reaffirms the climate change and the average global temperature increased by 0.74°C during 1906–2005, and it is expected to increase more in the future (see IPCC Reports, UNFCCC).

The rising temperature associated with increasing GHG in the atmosphere is interrelated to the global climate systems. Warmer temperatures are the cause of other major changes around the world. Climate change impacts include a rise in weather-related incidents such as floods, droughts, destructive storms, frosts and hailstones; the extinction of countless flora and fauna; the loss of agricultural crops in vulnerable areas; the changing of growing seasons; the melting of glaciers; the disruption of water supplies; the expansion of infectious diseases; the rising sea levels; and much more. Both the year 2011 and 2012 produced a record number of extreme climate events in the world including floods, heat waves, droughts, fires and snowstorms.

Climate change is a threat to this modern civilization and challenges to the developmental activities in this century. Truly, the 'climate change' is a by-product of industrialized nations – a result of accumulation of fossil fuel consumption in developed countries during industrialization which is the main cause of climate change in the world, today. However, there is a debate on country's contribution and cost sharing for mitigating climate change. Developed countries have contributed a lot to change the recent climate. Less developed countries (LDCs) contribute negligible or little to cause climate change (see UN Special report 2003, Khatun 2010, Coondoo and Dinda 2002, World Bank 1992, 2008), yet they face its harsh impacts and have the weakest capacity to adapt to these impacts (World Bank 2008).

1.1.2   Environmental Goods and Services

An environmental good can be understood as equipment, material or technology used to address a particular environmental problem or as a product that is itself 'environmentally preferable' to other similar products because of its relatively benign impact on environment. Environmental services are provided by ecosystems or human activities to address environmental problems and minimize the environmental damages and protect the biosphere of the Earth. Environmental goods and services (EGS) can be also classified as *environmental goods* comprising of pollution management products, cleaner technologies and products, heat and energy management, noise and vibration abatement, wastewater treatment equipment, resource management and environmentally preferable products; and it has also environmental services comprising of sewage services, refuse services, sanitation and similar and other services.

Climate Friendly Goods and Technology

Climate friendly goods and technologies (CFGT) is defined as components, products and technologies which have relatively less adverse impact on the environment (Dinda 2014a, b). The CFGT is a part of the wider group named environmental goods and services (EGS), which can be attributed to multiple-end use, relativism and like products at WTO (Balineau and de Melo 2011; Nguyen and Kalirajan 2015). EGS could be any equipment, material or technology which definitely must address certain environmental problem or *environmentally preferable* product which has relatively less harmful (or negative effect) on environment and human health. Environmental goods and services (EGS) actually is defined and used to measure, prevent, minimize or correct environmental damage (OECD: Eurostat 1999). In this context, CFGT is considered to be equipment, or machine, technology or material for environmental management, or environmentally preferable product to similar products. It consists of articles of iron, aluminium, machinery and mechanical appliances, electrical machinery equipment, ships, boats and floating structures, glass and glassware articles, etc. CFGT constitutes low-carbon growth technologies. One of the sub-categories of CFGT is *energy-efficient lighting* that aims to improve energy efficiency, and *solar photovoltaic system* targets to capture the solar energy. *Clean coal technology* focuses to reduce environmental impacts, including technologies of coal extraction, coal preparation and coal utilization. *Wind technology*, another sub-category of climate friendly goods and technologies, focuses on wind energy generation and is composed of three integral components: (i) the gear box, (ii) coupling and (iii) wind turbine.

Background of Climate Friendly Goods and Technologies in Trade

The environmental goods and services (EGS) were first discussed as part of the liberalizing agenda in the DOHA round of the multilateral trading round in 2001. The countries had wanted the tariff and non-tariff barriers to go down for trade of such EGS as this may lead to adoption of cleaner and cost-effective technologies by firms and country at large and possibly mitigate climate change and improve energy efficiency.

Liberalization has followed three routes, namely, (i) the list approach, (ii) project/integrated approach and (iii) request for offer approach. Environmental goods are always part of trade agenda but are subsumed within industrial or agricultural negotiations. Free and liberalized trade can make available such goods for countries which have no access to the CFGT or wherein domestic industries are unable to produce them in sufficient scale or at affordable prices. For exporters, additional market access can provide incentives to develop new products or technologies with less greenhouse gas emissions or less pollution that minimizes environmental damages. As a whole, global climate impact will definitely be reduced. These were discussed at the multilateral forums as countries wanted a smaller list to liberalize and wherein negotiations could be easier done than concentrating on the entire list of environmental goods. For example, WTO suggested a list of 153 goods for liberalization, and out of it, the World Bank identified 47 products, which are diverse products from wind turbines to solar panels to water-saving shower. Similarly, the OECD and ICTSD provided their lists. Following the WTO and UNESCAP, we finalized the list of CFGT, which is the core focus of this study and highlights the first decade of the twenty-first century for Asia.

Trade Value of EGS and CFGT

Most of the exporters of EGS are the developed nation, but some of the developing countries are also becoming important players in the heat and energy management equipment, noise and vibration abatement and environmental services like air pollution control and solid waste management (Jha 2008, 2009). Global EGS industry was worth $650 billion US dollars in 2008. Trade in EGS was estimated at amount of $65 billion US dollars. CFGT exports to the world were worth $38 billion US dollars out of total world exports of $1488 billion US dollars in 2008 with world export share of CFGT working out to be 2.5% in the year 2008. This export share of CFGT has varied between 2.3% in 2002 and 2.8% in 2009 in the precrisis period, while it rises in postcrisis period from nearly 3% in 2010 to 6.3% in 2016. World imports of CFGT were worth $38 billion US dollars out of total world imports of $1557 billion US dollars in 2008 with world import share of CFGT working out to be 2.4%, and this share has varied from 2.2% in 2002 to 2.7% in 2009 and from 2.9% in 2010 to 6.4% in 2016. A sudden downfall has been observed in aggregate export and import of CFGT following declining trend in global trade in the period of global financial crisis; however, after crisis, the volume and, more specifically, share of CFGT increase gradually over time. Our preliminary observations do suggest that CFGT trade in Asia has regional bias in the precrisis period and needs to be examined in detail.

1.1.3 Objective of the Study

The objective of this study is to (i) investigate trade performance of climate friendly goods and technologies (CFGT) in detail including its sub-groups; (ii) estimate its potential trade opportunity in Asia, and subregion like South Asia; and (iii) estimate India's position in South Asia and policy strategy on it. This study has several chapters with major focus on trade performance of CFGT and their classified product groups, competitiveness, regional orientation and estimation of potential trade gaps in India, South Asia and Asia. This study investigates potential trade opportunity of such goods in emerging India, South Asia and Asia which help to clarify Asia's position on global warming and regional efforts to mitigate climate change.

1.1.4 Study Plan

The book is organized as the following: Chap. 2 reviews the existing literature focusing on trade gravity models. Chapter 3 describes data and methodology. Part II provides trade performance briefly highlighting export, import, RCA and competitiveness. Chapters 4, 5 and 6 cover all these. Chapter 7 shows the regional orientation in Part III. Part IV analyses empirical issues. Chapters 8, 9 and 10 estimate and analyse the gravity model in Asia, South Asia and India, respectively. Chapters 8, 9 and 10 quantify the gap between expected and actual trade and contribute in the empirical measurement of potential trade gap of climate friendly goods and technologies in Asia, South Asia and India, respectively. Chapter 11 traces the direction of movement of CFGT and potential trade partners, and finally, Chapter 12 concludes with remarks.

Chapter 2
Literature Review

Abstract This chapter reviews the existing literature focusing on background of trade debate, environmental degradation and climate issues. It emphasizes on history of trade gravity model and highlights major empirical trade gravity findings in developed and developing countries.

Keywords EKC · NAFTA · Kyoto agreement · Technology transfer · Trade diversity · Climate change · Environmental degradation · EGS · CFGT · Potential trade · Doha round · Trade gravity model · Gravity equation · RTA · Efficient technology · Cross-border trade · Clean product · Energy-efficient technology · Asian financial crisis · Global financial crisis · Emerging market

This chapter briefly reviews available literature on trade and climate change. The pioneering work of Grossman and Krueger (1991) has evoked considerable discussion on the linkage between environmental quality and economic growth in the *North American Free Trade Agreement (NAFTA)* zone. Economic growth through industrialization generates more and more income, which acts as a *magnifier* of environmental degradation (Dinda 2004). Several studies (Grossman and Krueger 1995, World Bank 1992, Selden and Song 1994, Dinda 2004, and Stern 2004, etc.) have documented a systematic inverted U-shaped relationship between environmental degradation and income, which has been termed as the *environmental Kuznets curve (EKC)*. Trade is the most important factor that explains the EKC (Dinda 2004). Economic growth degrades environmental quality through three different channels, viz. scale, technology and composition effects (Copeland and Taylor 2004). Environmental quality could decline through the *scale effect* as increasing trade volume (especially export). Thus, trade might be a cause of environmental degradation, ceteris paribus. Trade may be good for environment as well through the *composition* and *technological effects*. As income rises through trade, environmental regulation is tightened; as a result, *pollution-reducing innovation gets promoted* (Porter and Linde 1995).

Trade plays a major role in innovations and disseminating technologies. Liberalized trade is a potent driver for technological innovation. Advanced know-how and environment-friendly technologies will be readily available through liberalized trade (World Bank 2008, Meyer-Ohlendorf and Gerstetter, 2009). So, trade liberalization is good for the environment (Antweiler et al. 2001, Liddle 2001, Copeland and Taylor 2004, Blyde 2000). Free trade has a contradictory impact on the environment, both increasing pollution and motivating reduction of it. Trade[1] can help developing countries adapt to generate the export earnings and access updated technologies (World Bank 2008). Trade has a role in the mitigation of climate change through disseminating and exchanging low-carbon technologies, which improves energy efficiency and reduces environmental impact.

Liberalized trade can provide or make available climate friendly goods and technology for countries which have no access to it or where domestic industries are unable to produce them in sufficient scale or at affordable prices (Dinda 2011a, b, 2014a, b). Through trade, additional market access may be possible that can provide incentives to exporters to develop new products or technologies which have less adverse impact on the environment. Most of the exporters of CFGT are from developed nations, but a few exporters are from developing countries.[2] After Kyoto agreement in 1997, some developing nations are also becoming important players in heat and energy management equipment, in noise and vibration abatement and in environmental services like air pollution control and solid waste management.[3] Few developing countries are among the top ten importers and exporters in various categories of CFGT which are relevant for adaptation and climate change mitigation (World Bank 2008). Climate change issues truly provide an opportunity to redesign economic activities including trade diversification focusing on CFGT (UNESCAP 2010, 2011).

There is a growing trend among industries to reconsider their production processes, thereby taking the environmental consequences of production into account. This concerns not only traditional technological aspects but also the organization of production and the design of products. Technological changes associated with the production process would also result in changes in the input-mix of materials and fuels, which may result in lower environmental impacts (Dinda 2004). Most of the developing countries rely on technology transfer through foreign direct investment from developed countries as a primary means of technology acquisition.

In this context, developing countries must focus on the production and trade of CFGT and need more emphasis on it. Production structure of the economy tends to shift towards cleaner activities which generate lesser pollution. Only few develop-

---

[1] Trade raises income level in developing economies, and it will create demands for tighter environmental protection, but lower trade barriers could hurt environment if heavy polluters move to countries with weaker regulations (Dinda 2004, Mukhopadhyay and Chakraborty 2005, Dean et al. 2009).

[2] Only Brazil and Mexico are producers of biofuels, and China is the exporter of energy-efficient lightings.

[3] See Veena Jha (2008) for more details.

ing nations, such as Brazil, China and Mexico, have already started to produce and emerge as important producers of clean energy technologies, while major developing countries are net importers (World Bank 2008).

2.1   Trade Debates on Environment and Climate Change

There is debate on trade resistances that might limit or promote trade between particular trading partners, often relying on a number of variables to proxy total trade resistances, including trade-related costs. Recently, global climate change itself creates new resistances on international trade after failure of the Doha round. This climate resistance also creates the opportunity for trade in new direction in the name of climate friendly goods and technology (CFGT) or green business opportunity.

2.2   Estimating Potential Trade

Literature (Baldwin 1994, Nilsson 2000, Egger 2002, Dinda 2011a, b, 2014a, b, etc.) uses the term *trade potential* as the expected volume of trade between country pairs that the gravity model predicts. Literature provides a measure of performance of bilateral trade flow and how well it performs relative to the model-predicted mean value. Literature measures how far above or below potential trade is from actual trade. Following the standard gravity model, Dinda (2014a, b) investigates a new direction of potential trade opportunity for environment-friendly goods and provides certain insights regarding trade opportunity of CFGT in Asia.

The trade gravity model is based on the idea that trade volumes between two countries depend on the size of the two countries and the distance between them. Distance between pair of nations can be geographical, cultural and political. Socio-economic-political and cultural aspects may create obstacle to adopt updated cleaner technology in certain societies or countries. Trade literature addresses these socio-economic and cultural issues in the empirical investigations. Eichengreen and Irwin (1998) and Rauch (1999) demonstrate cultural proxies (border, common language) as dummy variables in the empirical gravity equation. Geographical distance is the proxy of the cost of transport. The gravity model has been used extensively in analysing and explaining trade. Harrigan (2001) and Anderson and van Wincoop (2004) review comprehensively on trade resistances which might limit or promote trade between trading partners, trade resistances and costs of trade. Recently, the failure of incorporation of climate change in the Doha round might create new resistances on international trade. On other hand, this climate resistance may create opportunity for trade in new directions like CFGT trade or green businesses. The review of literature demonstrates the new direction of potential trade in CFGT.

The gravity model is clearly distinguished and tractable well representation of economic interaction in a multi-country world. Traditional trade theory is concentrated on two or three country cases focusing on special features. The distribution of goods or factors across space is determined by gravity forces given the size of economic activities for each and every location. Gravity model is the most successful empirical analysis in applied economics for understanding the trade relations and the distribution of goods or factors of production. Gravity model provides more accurate estimation and predicts potential trade.

2.3   Literature on Trade Gravity Model and Its Applications

Ravenstein (1889) was the pioneer for the use of gravity model for migration patterns in the UK in the nineteenth century. Tinbergen (1962) was the first who used gravity model to explain trade flows. Anderson (1979) introduced the gravity model theoretical legitimacy and popularized it in trade empirics. Tinbergen (1962) points out empirically that international trade between two countries is determined by their relative masses and their distance from each other. The gravity model has been used extensively in empirical international trade since its inception. Over time this model has been used largely in explaining the effects of different policies and other determinants of trade flows, with key variables of economic size and distance. Its popularity in empirical research has increased rapidly with the introduction of theoretical gravity by Anderson (1979). Anderson (1979) derived the gravity equation from expenditure systems where goods are differentiated by country of origin and distance is the proxy of all transport costs. With the assumption of frictionless trade, Anderson (1979), Helpman (1987) and Deardorff (1995) describe the theoretical foundations of the gravity model. They derive a model where trade volumes between country pairs are proportions of the product of incomes or total world trade. Bergstrand (1985) provides a theoretical underpinning and derivation of the model as a 'partial equilibrium subsystem of a general equilibrium model'. Helpman (1987) derives the gravity model from an imperfect competition model, and Deardorff (1995) derives it from the Heckscher-Ohlin model. Indeed, the gravity model can be derived from numerous trade theories in one form or another and can be used to find empirical evidence of many trade theories with different assumptions about preferences and whether goods are differentiated or homogeneous (Deardorff 1995; Harrigan 2001). Trade shares 'fall naturally into a gravity-equation' (Deardorff 1995). The probabilistic method is comparable to the analysis of trade intensities (Drysdale and Garnaut 1982) which uses the relative size of a country's trade as a benchmark for what the country is expected to trade. Their theoretical gravity equation is based on the assumptions of frictionless trade or iceberg transport costs to capture all the frictions.

Linnemann (1966) started a process in the literature of adding trade explicators and inhibitors to the gravity model. Using the gravity model as the main tool, Frankel, Stein and Wei (1997) undertake a comprehensive study of regional trading blocs. Frankel, Stein and Wei (1997) comprehensively convinced both discriminatory and non-discriminatory effects of bilateral and trade arrangements. They are able to quantify the amount by which different preferential trade arrangements (PTAs) and regional arrangements such as APEC increase trade by adding trade agreement dummy variables into the standard gravity model. Using gravity models, analysis of regional or multilateral trade arrangements is now common and important in applied trade theory. Rose (2000) made an important contribution as the first to include a common currency dummy variable to explain trade. The finding that an economy which is so highly integrated with another economy that there is a common currency increases trade threefold, as his European Union dummy suggested, had a large impact on the literature with significant policy implications. The idea of increased trade from a common currency is intuitive, but the magnitude was surprising. Baldwin and Taglioni (2006) reduce the magnitude of the common currency effect significantly using Anderson and van Wincoop (2003)'s structural estimation with multilateral resistance. McCallum (1995) found that trade between the USA and Canada was lower than trade within their borders by a factor, but Anderson and van Wincoop (2003) reduce this unexplained border effect to the border's lowering trade by 44 per cent. They assumed symmetric trade costs to solve their model, which is a significant but unrealistic assumption. Relaxing symmetric border cost assumption, Balistreri and Hillberry (2006) account for structural bias in Anderson and van Wincoop (2003) that arises due to incorrect treatment of an adding-up constraint which is implicit in the Anderson and van Wincoop (2003) model. The correct estimation of the Anderson and van Wincoop (2003) derivation shows that the literature still cannot explain the border puzzle or what we prefer to describe here as unexplained resistances. Anderson and van Wincoop (2003) claim to solve the border puzzle using McCallum's data by deriving the gravity equation from expenditure functions and importantly adding what they call multilateral resistance.[4]

McCallum (1995) applied the gravity model to estimate a value for the loss in trade volume accounted for by goods crossing the USA-Canada border as compared to intranational trade in both countries and prefers to describe as unexplained resistances. The findings show that international border effects are inferred and that they matter even with two economies that share a large border and are highly integrated through a regional trade arrangement (RTA) such as NAFTA. Trading across borders will cause disconnect in relative prices as insurance, freight, tariffs and non-tariff barriers and different regulatory structures cause uncertainty and impede trade to some extent.

The wide use of the gravity model and the policy implications drawn from its applications are quite significant in literature on the accuracy of the econometric specifications and techniques. Different econometric specifications of gravity

---

[4]The multilateral resistance terms are important and mean that if country i's trade with country j is being analysed and there is no movement in the trade determinants, a change in country k's trade with country i will affect the trade between i and j, as would be expected. Their specification explains away most of the border puzzle.

equations are used in literature. The question of using population as an explanatory variable is one example where the gravity equation is inconsistent. Anderson (1979), Helpman (1987) and Deardorff (1995) do not justify the inclusion of population; its effect could be positive or negative. A positive effect would be the expected result for developing economies having higher population as they tend to be specialized in labour-intensive exports and trade more. A negative effect for population size could be due to economies with larger populations having an absorption effect (Martínez-Zarzoso and Nowak-Lehmann 2003). Including the log of GDP and log of population separately, the log linearization of the gravity model for estimation is equivalent to including the log of GDP per capita with a restriction on the estimated coefficients of GDP and population separately. The reason GDP per capita is included in so many models is that it has meaning in the context of using the Linder hypothesis in explaining trade flows. Baldwin and Taglioni (2006) summarize errors that are frequently repeated in the literature.

Chapter 3
Methodology and Data

Abstract This chapter describes data which is used in this study. It discusses methodologies that are applied in this study. Several trade indices are defined with their properties.

Keywords Competitiveness index · Michelaye index · Regional orientation index · Trade gravity model · CFGT · UN COMTRADE · UNESCAP · The World Bank · OECD · WTO · Potential opportunity · Efficient technology · CCT · WE · SPVS · EEL · Asia · Emerging

3.1 Methodology

Appropriate measurement tools are used to judge the trade performance, and gravity technique is applied to measure and estimate trade gaps. This chapter discusses these.

Trade performances of countries or group of nations are judged using some trade indices and indicators. Trade indices like export and import shares, revealed comparative advantage index, competitiveness index, regional orientation index and Michelaye index for trade of CFGT and its sub-categories for Asian nations and other regional groups are calculated to form a policy opinion on countries' competitiveness, trade patterns, changing comparative advantage over time and regional bias.

3.1.1   Competitiveness Index

Competitiveness index (CI) measures a country's export share in the world export. It is defined as

$$CI = \frac{\sum_n X_p}{\sum_w X_p} * 100,$$

where p is the product, n is country and w is world. The value of competitiveness index lies between 0 and 100. Competitiveness index truly indicates market share of a country which also reflects its market power and control.

3.1.2   Revealed Comparative Advantage

The concept of revealed comparative advantage (Balassa 1965, 1977, 1979, 1986) pertains to the relative trade performance of individual countries in particular commodities. On the assumption that the commodity pattern of trade reflects intercountry differences in relative costs as well as in non-price factors, this is assumed to "reveal" the comparative advantage of the trading countries.

Following Balassa (1965) this study measures reveal comparative advantage (RCA) of a product of a country. RCA is the ratio of a country's share of world exports of a product to its share of total world exports of all products. Revealed comparative advantage of jth product of ith country (RCA_{ij}) is defined as

$$RCA_{ij} = \left(\frac{X_{ij}}{X_{wj}}\right) / \left(\frac{X_i}{X_w}\right)$$

where X_{ij} = ith country's export of jth commodity,

X_{wj} = world total exports of jth commodity,

X_i = total exports of ith country, and

X_w = total world exports of all commodities

The RCA is measured using post-trade data. The index of revealed comparative advantage (RCA_{ij}) has a relatively simple interpretation. If it takes a value greater than unity, the country has a revealed comparative advantage in that product. The advantage of using the comparative advantage index is that it considers the intrinsic advantage of a particular export commodity and is consistent with changes in an economy's relative factor endowment and productivity. The disadvantage, however, is that it cannot distinguish improvements in factor endowments and pursuit of appropriate trade policies by a country.

3.1.3   Michelaye Index

Michelaye index is defined as the difference between the share of a country's total exports of a product in its total exports and the share of the same country's imports of the same product in its total imports.

$$\text{Michelaye Index} = \frac{\sum_n X_p}{\sum_n X_{pi}} - \frac{\sum_n M_p}{\sum_n M_{pi}}$$

where p is a specific product, n is country, pi is all products, X is exports to the world and M is imports from the world. The first term is the share of export of product p in total export of country n, and second term is the share of import of product p of country n. It compares the export pattern of a country to its own import pattern. The value of Michelaye index ranges from -1 to $+1$.

3.1.4   Regional Orientation

Regional orientation index (ROI) is the ratio of two shares. The numerator is the share of a country's exports of a given product to the region of interest in total exports to the region. The denominator is the share of exports of the product to other countries in total exports to other countries. It is defined as

$$\text{ROI} = \frac{\sum_r x_{kir} / \sum_r X_{ir}}{\sum_w x_{kiw} / \sum_w X_{iw}}$$

where i is the country of interest, r is the set of countries in the regional block, w is the set of all countries not in the bloc, k is the sector of interest, x is the commodity export flow and X is the total export flow. The numerator is the share of good k in the exports of country i to region r, while the denominator is the share of good k in the exports of country i to non-members of r.

ROI takes a value between 0 and $+\infty$. A value greater than unity (ROI > 1) implies a regional bias[1] in exports. The ROI indicates whether exports of a particular product from one region under study to a given destination are greater than exports of the same product to other destinations. ROI measures the importance of intraregional exports relative to outside regional exports.

---

[1] Limitations: The index may be affected by many factors, including geographical ones. Because it is based on relative shares, a strong regional orientation may be of little economic significance.

3.1.5 Gravity Model

The traditional trade gravity model is drawn on an analogy with Newton's law of gravitation, which states that the gravity between two objects is directly related to their masses and inversely related to the distance between them. Now, one country could be perceived as object having economic factors or resources attracting mass economic activities. A mass of goods or factors of production supplied at origin i, Y_i, is attracted to a mass of demand for goods or factors of production at destination j, Y_j, but the potential flow is reduced by the distance between them, D_{ij}. Strictly applying the analogy,

$$X_{ij} = \frac{Y_i Y_j}{D_{ij}^2} \tag{3.1}$$

Equation (3.1) provides a definite and the predicted trade movement of goods or factors of production between country i and country j, X_{ij}. The trade gravity model is initially presented as an intuitive way of understanding trade flows. Trade values or volumes are unpredicted due to uncertainty which is associated with randomness. Truly, for economic analysis a stochastic variable might be associated with the gravity model. Adding one stochastic variable (ζ) to Eq. (3.1), and then, Eq. (3.1) turns to be

$$X_{ij} = \frac{Y_i Y_j}{D_{ij}^2} \zeta_{ij} \tag{3.2}$$

The Eq. (3.2) is the stochastic gravity model which is applicable in econometric analysis. Now, rearranging Eq. (3.2) in the product form, the stochastic version of the gravity model (Eq. 3.2) could be expressed in the productive form with parameters. So, the stochastic version of the gravity model (2) in productive form is formulated as

$$X_{ij} = \alpha Y_i^{\beta_1} Y_j^{\beta_2} D_{ij}^{\beta_3} \xi_{ij} \tag{3.3}$$

where $\alpha, \beta_1, \beta_2, \beta_3$ are unknown parameters ($\beta_3 < 0$ and it is associated with D in Eq. (3.2)) and ξ_{ij} is the stochastic term. Truly, D is physical distance between objects; however, in economics, this distance, D, could be geographical or physical distance, political distance and cultural distance.

Taking log of both sides of Eq. (3.3), it will be log linear equation as

$$\ln(X_{ij}) = \ln(\alpha) + \beta_1 \ln(Y_i) + \beta_2 \ln(Y_j) + \beta_3 \ln(D_{ij}) + \ln(\xi_{ij}) \tag{3.4}$$

In its basic form, the gravity model can be written as

$$\ln\left(X_{ij}\right) = \beta_0 + \beta_1 \ln\left(Y_i\right) + \beta_2 \ln\left(Y_j\right) + \beta_3 \ln\left(\tau_{ij}\right) + \varepsilon_{ij} \tag{3.5}$$

where X_{ij} indicates exports from reporting country i to trading partner country j; $\beta_0 = \ln(\alpha)$; Y is each country's gross domestic products (GDP); $\ln(\tau_{ij}) = \ln(D_{ij}) \equiv \ln(\text{Distance}_{ij})$ or τ_{ij} represents trade costs between two countries; distance is the geographical distance between them – as an observable proxy for trade costs; and ε_{ij} is a random error term.

Tinbergen (1962) introduced the gravity model to explain trade flows, while Anderson (1979) popularized the traditional gravity equation 3.5). However, the popularity of the gravity model increased due to Anderson and Wincoop (2003), and it has now become the de facto standard in empirical work. Contiguity or common border, common language, common ethno group, colony, small country, etc., are used as dummy variables and added in the basic gravity Eq. (3.5). For the estimation and analysis purpose, the following equation is considered for the study purpose:

$$\ln X_{ij} = \beta_0 + \beta_1 \ln GDP_i + \beta_2 \ln GDP_j + \beta_3 \ln PCGDP_i + \beta_4 \ln PCGDP_j$$
$$+ \beta_5 \ln DT_{ij} + \beta_6 D_{contig} + \beta_7 D_{comlang} + \beta_8 D_{comlang\_ethno} + \beta_9 D_{colony} \tag{3.6}$$
$$+ \beta_{10} D_{comcol} + \beta_{11} D_{col45} + \beta_{12} D_{smctry} + \beta_{13} Trf_j' + \varepsilon_{ij}$$

where X_{ij} denotes the value of country i exports to country j; GDP_i and $PCGDP_i$ denote the exporting country's gross domestic product and per capita income (Markusen 2013), respectively; GDP_j and $PCGDP_j$ denote the gross domestic product and per capita income[2] of the partner of the exporting country, respectively; DT_{ij} denotes the distance between reporting country and its partner country; Trf_j is the (weighted average) tariff rate imposed by the partner of the exporting country; and $D_{contig}, D_{comlang}, D_{comlang\_ethno}, D_{colony}, D_{comcol}, D_{col45}$ and D_{smctry} are the dummy variables for contiguity, common language, colony, common colony, colony from 1945 and small country, respectively. One country's economic activity is reflected in its income level or GDP, and development position is observed in its per capita GDP level. In our regression analysis, we have used the log values of all the variables[3] except for dummies. It reduces the heteroscedasticity problem to some extent. For the analysis purpose, the least square econometric technique is used to estimate the above gravity Eq. (3.6) and find predicted trade for each country. Trade gap is measured by the estimated bilateral residual. Truly, trade gap is the difference between actual and model predicted trade. There are a lot of other unknown factors restricting CFGT trade flows. So, trade opportunity may rise with reducing restrictions.

---

[2] Per capita income is the proxy of level of economic development (Markusen 2013).

[3] In original version of Tinbergen (1962), the model is expressed in a log-log form. So the parameters are elasticity of the trade flow with respect to the explanatory variables.

3.2 Data Description

The environmental goods and services (EGS) is a broader group of environmental-friendly products, while climate friendly goods and technologies (CFGT) are a sub-group and form a part of the broader group of EGS. These were discussed at the international levels and/or multilateral forums. There was no consciousness identifying EGS. Most of the countries wanted a smaller list to liberalize at initial stage with less confrontation in the negotiation table. This study has selected 64 climate friendly goods under 6-digit HS code (2002) by putting together various lists that have been defined by various international organizations recently. In the global platforms wherein negotiations could be easier done than concentrating on the entire list of environmental goods. For example, the WTO provided a list of 153 goods for liberalization in international trade arena. The World Bank identified and selected 47 products out of 153 products list proposed by proponents of Environment Goods liberalization in the WTO. These 47 products comprised diverse products from wind turbines to solar photovoltaic system and activated carbon to water-saving shower. Similarly OECD and ICTSD had their own lists of environmental goods and services. The list is arrived by defining concordance series from series of lists given by the World Bank, ICTSD, WTO, APEC and the OECD. Climate friendly goods and technologies (CFGT) trade data (in value, 1000 US dollar at 2008) was taken from the WITS (World Integrated Trade Solution) website: https://wits.worldbank.org having the UN COMTRADE data for the period of 2002–2017. Gross domestic production (GDP) and per capita GDP data were taken from the World Bank Development Indicators (www.worldbank.org/data) for corresponding years. The distance between countries and other dummy variables are taken from the dist_cepii.xls file of CEPII database (see the website: www.cepii.fr). The total observation was reduced after combining all the variables for each pair of trading partners.[4] After matching all data sets, this filtered data set is used in the empirical analysis.

This study has been able to identify 64 climate friendly goods and technologies (CFGT) under 6-digit HS code (2002). Various international organizations recently define and identify CFGT. The list was prepared from a series of lists given by the World Bank, ICTSD, WTO, APEC and the OECD. This study considers these 64 CFGT as one category and estimates the above-mentioned trade indicators. Appendix I provides the entire list of the 64 CFGT.

Following UNESCAP and the World Bank (2008), this study also divides these CFGT into (i) clean coal technologies (CCT), [HS code 840510, 841181 and 841,182]; (ii) wind energy (WE), [HS code 848340 and 848,360]; (iii) solar photovoltaic systems (SPVS), [HS code 850720, 853710 and 854,140]; and (iv) energy-efficient lighting (EEL), [HS code 853931]. Besides these four sub-groups, the paper has also considered the fifth group as 'other codes' (OC) that consists of all HS codes not considered in the abovesaid four sub-categories.

---

[4] This study considers fully matched data only.

Part II
Trade Performance

Chapter 4
Trade Performance of CFGT in Asia

Abstract This chapter provides trade performance of CFGT in Asia. Average growth rate of CFGT in Asia was around 8% in the period of 2002–2017. It also highlights more on trade of sub-categories in the pre- and post-global financial crisis. Japan was the top exporter of wind energy and solar photovoltaic system in the pre-global crisis period; China emerged and replaced Japan in the post-global crisis era. Export share of SPVS in China increased over time during 2002–2008 and reached at 2nd position in Asia in 2009. Truly, exports of EEL and SPVS were concentrated in East Asia region since the beginning of the twenty-first century. Most of countries in Asia were CFGT importers from countries within Asia, the European Union, and the USA and Canada. Ten major developing economies of Asia had import shares above the world average, which indicate that Asia has emerged as powerhouse of the world.

Keywords Climate change · Export share · Import share · CFGT · Potential opportunity · CCT · EEL · WE · SPVS · SAARC · ASEAN · APTA · Emerging market · East Asia · Renewable energy

4.1 Introduction

This study, initially, examines trade performance of CFGT in Asia and highlights trade of sub-categories including global financial crisis in 2008–2009 in the period of 2002–2017. Selecting major CFGT trade items, we examine the trend of export and import of CFGT trade in Asia during 2002–2017. Overall trade of CFGT increased during 2002–2017 except a downfall in 2009 (Fig. 4.1). However, for the said period, on an average growth rate of export and import of CFGT were 8.1% and 8.3%, respectively.

We observe also the trends of sub-categories of CFGT trade in Fig. 4.2. Asia enjoyed comparative advantage in exporting EEL compared to the rest of the world. Trade share of SPVS in CFGT in Asia contributed maximum during 2002–2017. Considering all CFGT in this study, it is noted that imports of 64 CFGT dominated

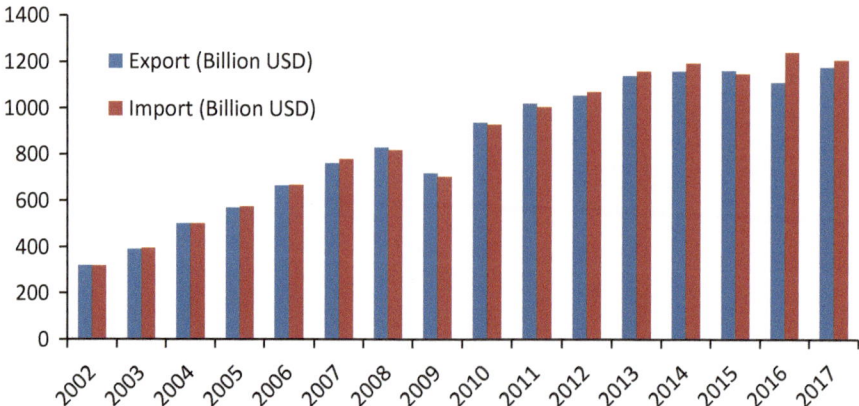

Fig. 4.1 Export and import of selected major CFGT items in Asia during 2002–2017

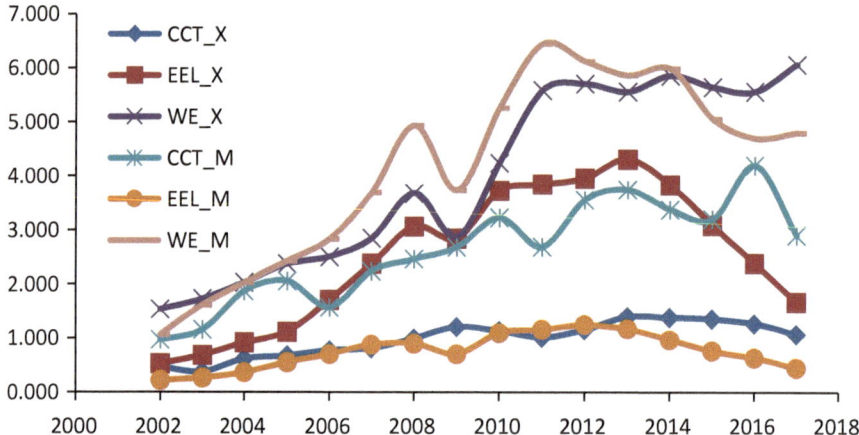

Fig. 4.2 Export and import of sub-categories of selected CFGT in Asia during 2002–2017

in Asia in precrisis period (2002–2008), while these exports in Asia increased at faster rate than import in postcrisis period (2010–2017). These preliminary observations motivate to investigate trade performance of CFGT in details in the precrisis period of 2002–2008. Trade diversity in CFGT, a new direction of trade, emerged in Asia from beginning of the twenty-first century.

This chapter assesses trade performance of CFGT in details using trade indices like export and import shares in Asia and its regional trade blocks like the South Asian Association for Regional Cooperation (SAARC), Asia-Pacific Trade Agreement (APTA) and Association of Southeast Asian Nations (ASEAN[1]) in the

---

[1]All countries of ASEAN and APTA fall under the wider Asia and Pacific region, which is spread over Russia to New Zealand.

precrisis period. This study also discusses some trade indices and/or indicators to judge CFGT trade performance in Asia and its subregional trading groups during 2002–2008.

4.2   Analysis of CFGT Trade Performance in Asia

4.2.1   Export Share of CFGT in Asia

Export share is the ratio of Asia's total exports of CFGT product to the World CFGT export to Asia's total exports of all products to the world. Following Dinda (2015) this study finds out the export share for countries in Asia for the period of 2002–2008. Table 4.1 shows export and import share of CFGT in Asia for 2002, 2005 and 2008. Table 4.1 is partitioned vertically into two parts displaying export and import and divided horizontally into three parts for 2002, 2005 and 2008. Left and right sides of Table 4.1 provide export- and import-related findings, respectively. Left side of Table 4.1 provides CFGT export share of reporting countries along with CFGT exports to the World, and their ranking is based on export share for 2002, 2005 and 2008; and right side of Table 4.1 displays its import share and corresponding ranks for the said years.

Japan's export share of 4.01% in 2002 is calculated by taking the ratio of CFGT exports to the world by Japan (1.67 billion USD) to exports of all products to the world by Japan (41.65 billion USD) and multiplied by hundred. The ratio was above the world export share of CFGT goods in 2002. Similarly, Hong Kong had export shares above the world average depicting its excellent trade performance for CFGT in 2002. The study finds that Japan and Hong Kong are performing better than the world average in 2002. In 2002, CFGT export share of South Korea, Singapore, Malaysia and Thailand were 2.06%, 1.65% and 1.59%, respectively.

Middle left part of Table 4.1 provides CFGT exports of reporting countries along with their export share and corresponding ranking based on export share for 2005. Japan's export share of 4.88% in 2005 is calculated using the ratio of CFGT exports to the world by Japan (2.9 billion USD) to exports of all products to the world by Japan (59.548 billion USD). Japan, Hong Kong and China have export shares above the world average depicting its excellent trade performance for CFGT in 2005. CFGT export shares of Hong Kong, China, South Korea and Thailand were 2.67%, 2.34%, 1.87% and 1.77%, respectively. India was placed in top ten Asia's CFGT exporter list in 2005 and emerged in CFGT trade in Asia.

Left of last part of Table 4.1 shows CFGT exports of reporting countries along with their export share to the world and corresponding ranking based on export share in 2008. Japan's export share of 5.20% in 2008 is calculated using the ratio of CFGT exports to the world by Japan (4.0664 billion USD) to exports of all products to the world by Japan (78.140 billion USD). The ratio was above the world export share of CFGT goods in 2008 of 2.5%. Similarly, China and Hong Kong had export

Table 4.1 CFGT export and import shares of top ten Asian countries in world export and import in 2002, 2005 and 2008

Exports in 2002				Import in 2002			
Countries	CFGT exports (million US$)	Export share (%)	Rank 2002	Countries	CFGT imports (million US$)	Import share (%)	Rank 2002
Japan	1670.89	4.01	1	China	1063.787	3.604	1
Hong Kong	516.725	2.56	2	Thailand	210.365	3.254	2
China	739.43	2.27	3	Turkey	166.735	3.252	3
South Korea	334.212	2.06	4	South Korea	451.0	2.965	4
Singapore	207.031	1.654	5	Malaysia	231.878	2.947	5
Malaysia	153.321	1.63	6	Singapore	318.187	2.73	6
Thailand	108.316	1.59	7	Russia	114.084	2.47	7
Turkey	38.466	1.076	8	Hong Kong	478.381	2.30	8
Sri Lanka	2.804	0.594	9	Macao	5.0557	1.998	9
Russia	60.421	0.566	10	Sri Lanka	11.73	1.943	10
Exports in 2005				**Import in 2005**			
Japan	2905.938	4.88	1	Iran	160.02	4.138	1
Hong Kong	779.614	2.67	2	Georgia	9.474	3.80	2
China	1783.053	2.34	3	Bhutan	1.4384	3.717	3
South Korea	531.905	1.87	4	Kazakhstan	62.372	3.594	4
Thailand	194.84	1.77	5	China	2364.956	3.583	5
Malaysia	228.811	1.62	6	Maldives	2.669	3.583	6
Singapore	300.80	1.31	7	South Korea	877.215	3.358	7
Turkey	91.475	1.25	8	Thailand	374.44	3.169	8
India	101.572	1.01	9	Azerbaijan	13.32	3.163	9
Macao	1.8876	0.76	10	Russia	282.917	2.866	10
Exports in 2008				**Import in 2008**			
Japan	4066.4	5.204	1	Kazakhstan	155.177	4.104	1
China	4885.1	3.415	2	South Korea	1699.418	3.904	2
Hong Kong	977.71	2.64	3	Azerbaijan	27.632	3.858	3
South Korea	1015	2.405	4	China	3779.723	3.337	4
Philippines	114.34	2.33	5	Vietnam	265.491	3.289	5
India	354.98	1.95	6	Pakistan	118.693	2.804	6
Thailand	298.72	1.698	7	Thailand	494.262	2.767	7
Malaysia	316.29	1.590	8	Russia	691.527	2.59	8
Singapore	525.02	1.55	9	Hong Kong	983.969	2.504	9
Turkey	178.78	1.355	10	Malaysia	358.846	2.297	10

Source: Author's calculations

shares above the world average depicting excellent trade performance of such countries for CFGT in 2008. It should be noted that Japan, China, South Korea and Hong Kong were top four exporters in CFGT in Asia in the said period and seven out of ten CFGT exporting nations belonged to East and Southeast Asian regions.

4.2.2 Import Share of CFGT in Asia

Import share is the ratio of Asia's total imports of CFGT product to the world CFGT import to Asia's total imports of all products to the world. The study finds out the import share for countries in Asia in the period of 2002–2008. Top, middle and lower parts of right side of Table 4.1 display import shares of CFGT of reporting countries in Asia in 2002, 2005 and 2008, respectively. China, Thailand, Turkey, South Korea, Malaysia and Singapore were the top six countries in Asia having CFGT import share above 2.5% in 2002. Iran, Georgia, Bhutan, Kazakhstan, China, Maldives, South Korea, Thailand, Azerbaijan and Russia became the top ten Asian countries having CFGT import share above 2.5% in 2005. Nine of them had import share more than 3% in 2005 except Russia (2.866%). The top eight countries in Asia having CFGT import share above 2.5% in 2008 were Kazakhstan, South Korea, Azerbaijan, China, Vietnam, Pakistan, Thailand and Russia.

The findings show that countries with ranks 1 to 10 in both 2002 and 2008 have import shares above the world averages. China, Hong Kong, Thailand, South Korea, Malaysia, Singapore, Turkey and Russia were eight major countries in Asia above world import share of CFGT of 2.2% in 2002. Kazakhstan, South Korea, Azerbaijan, China, Vietnam, Pakistan, Thailand, Russia and Hong Kong were nine major Asian countries that crossed over the world import average of CFGT of 2.4% in 2008. These findings suggest that most of the countries in Asia were basically importers of CFGT products in our study periods. This study confirms the above statement following regional group performance of ASEAN, APTA and SAARC.

Japan was the top performer in CFGT export share during 2002–2008. Japan's export share of CFGT was 4.01% in 2002, and it increased from 4.88% in 2005 to 5.2% in 2008 (see Table 4.1). Japan, Hong Kong, China and South Korea were in top four ranks in Asia's CFGT export in the period of 2002–2008. Hong Kong's CFGT export share increased from 2.56% in 2002 to 2.67% in 2005, and it declined to 2.64% in 2008 while China's export share increased from 2.27% in 2002 to 2.34% in 2005 to 3.41% in 2008. CFGT export share of South Korea increased from 2.06% in 2002 to 1.87% in 2005 to 2.4% in 2008. CFGT export shares of Thailand and Turkey increased marginally while that of Singapore and Malaysia declined in 2008 compared to 2002. India has emerged and claimed a position in top ten Asia's CFGT export share 1.01% in 2005 and 1.95% in 2008. It should be noted that Singapore and Malaysia had been replaced with the Philippines and India for fifth and sixth positions in 2008 from the 2002 position. South Korea and Thailand were ranked fourth and seventh, respectively, in 2002 and 2008. The share of CFGT in

Table 4.2 Export and import shares of CFGT in world for subregional groups in Asia and its surroundings in 2002, 2005 and 2008

Region	Export share (%)	Rank	Region	Import share (%)	Rank
2002					
APTA	2.16	1	APTA	3.32	1
ASEAN	1.63	2	ASEAN	2.93	2
SAARC	0.32	3	SAARC	1.48	3
2005					
APTA	2.08	1	APTA	3.23	1
ASEAN	1.45	2	ASEAN	2.4	2
SAARC	0.86	3	SAARC	1.56	3
2008					
APTA	3.06	1	APTA	3.18	1
SAARC	1.73	2	SAARC	1.8	2
ASEAN	1.58	3	ASEAN	2.34	3

Source: Author's calculations

world exports increased for all countries in Asia. However, top performers in Asia's CFGT import change during 2002–2008.

The study also found out the export share for some regional groups for the period of 2002–2008. Table 4.2 provides export and import shares of CFGT in the world for subregional groups in Asia and its surroundings in 2002, 2005 and 2008. APTA region as group's export shares were more than the world average depicting the relatively better performance in comparison with other groups in the region, namely, SAARC and ASEAN. The share of CFGT in world exports increased in Asia, Southeast Asia, and South Asia regions from 2002 to 2008. There was a clear realization of cleaner technologies and goods trade in Asia and South Asia regions. The above were indications that most of the countries in Asia were basically importers of CFGT products from countries within Asia. This study looks at the regional group performance of ASEAN, APTA and South Asia and confirms the above statement.

APTA and ASEAN had import shares above world import share of CFGT in 2002, 2005 and 2008. In South Asia the export share of CFGT had increased more than that of import share during 2002–2008. Export share of CFGT was 0.32% in 2002, 0.86% in 2005 and raised it to 1.73% in 2008 while import share was 1.48% in 2002, 1.56% in 2005 and marginally increased to 1.8% in 2008 in South Asia region.

In APTA region, import share of CFGT declined from 3.32% in 2002 to 3.23% in 2005 to 3.18% in 2008, while its export share increased from 2.16% in 2002 to 2.08% in 2005 to 3.06% in 2008. Both export and import shares of CFGT in ASEAN region declined marginally during 2002–2008. For more detailed understanding, we have to investigate the change of trade pattern of sub-groups of CFGT post-Asian crisis to pre-global crisis (during 2002–2008).

Table 4.3 displays the trends of Asia's export shares of CCT, EEL, SPVS, WE and OC of CFGT in the period of 2002–2009 in columns 2–6, and import shares of the said sub-categories are shown in columns 7–11, respectively. Among CFGT, export share of CCT is the least in Asia. Export share of OC is the maximum and dominates Asia's CFGT export during 2002–2009. Both export and import shares of CCT and OC declined over time. SPVS and EEL export shares in CFGT grew over the period of 2002–2009, while export shares of CCT, WE and OC declined during 2002–2009 (see Table 4.3). Both export shares of SPVS (28.24%) and WE (4.37%) reached in peak in global crisis in 2009.

Trend pattern of import shares of all sub-categories were similar to export share trend except EEL. Import share of EEL of CFGT in Asia was the least and remained less than one per cent of CFGT for the said study period. Out of CFGT, OC holds maximum share both in export and import (see Table 4.3). It should be noted that in our study period import share CCT reached in peak in 2005, while that of EEL and WE attained peak in 2006 and 2008, respectively.

Now we also calculate average growth rate of export and import of CFGT and its sub-categories for the period of 2002–2008 (excluding trade global crisis), 2002–2009 (including trade in global crisis year) and 2002–2017. Table 4.4 shows the average growth rates of export and import of CCT, EEL, SPVS, WE, OC and CFGT

Table 4.3 Trends of Asia's export and import shares of CCT, EEL, SPVS, WE and OC of CFGT in the world trade in the period of 2002–2009, respectively

	Export share (%) of sub-categories of CFGT					Import share (%) of sub-categories of CFGT				
Year	CCT	EEL	SPVS	WE	OC	CCT	EEL	SPVS	WE	OC
2002	1.16	1.32	18.04	3.85	75.63	2.42	0.58	15.05	2.86	79.09
2003	0.80	1.40	20.38	3.59	73.83	2.34	0.56	15.31	3.38	78.40
2004	0.93	1.37	20.51	3.09	74.10	2.79	0.56	15.17	3.17	78.32
2005	0.94	1.49	20.71	3.23	73.63	2.83	0.68	15.46	3.37	77.66
2006	0.91	1.96	20.88	2.91	73.33	1.87	0.81	15.70	3.47	78.15
2007	0.73	2.19	20.15	2.56	74.37	1.86	0.75	14.88	3.44	79.08
2008	0.76	2.32	24.08	2.78	70.06	1.57	0.66	16.58	4.04	77.15
2009	0.33	4.37	28.24	1.67	65.38	1.88	0.59	18.33	3.74	75.46

Source: Author's calculations

Table 4.4 Average growth rate of CFGT and its sub-categories during 2002–2008 (excluding global crisis), 2002–2009 (including global crisis), and 2002–2017

	Trade	CCT	EEL	SPVS	WE	OC	CFGT
Avg. growth rate during 2002–2008	Export	13.2	29.6	25.0	14.7	18.8	20.1
	Import	12.3	21.7	21.1	25.2	18.9	19.3
Avg. growth rate during 2002–2009	Export	−1.2	23.7	12.9	−5.4	4.4	6.5
	Import	4.9	8.8	11.3	12.3	7.7	8.4
Avg. growth rate during 2002–2017	Export	5.2	7.2	8.2	8.6	7.6	8.2
	Import	6.9	4.5	8.4	9.4	8.3	8.4

Source: Author's calculations

for the period of 2002–2008, 2002–2009 and 2002–2017. Average growth rates of CFGT export were 20.1%, 6.5% and 8.2% in the period of 2002–2008, 2002–2009 and 2002–2017, respectively; and that of CFGT import growth declined from 19.3% in the period of 2002–2008 to 8.4% in the period of 2002–2009, and overall average growth rate of CFGT import was 8.4% during 2002–2017.

The highest export growth rate was observed in EEL in Asia and the second highest growth rate in SPVS during 2002–2009. Average growth rate of EEL import declined from 21.7% to 8.8% due to global crisis; however, that of EEL export reduced to 23.7% from 29.6%. Both export and import growth rates of SPVS became nearly half for global crisis. Import growth rate in WE was the highest in Asia in our study periods.

Table 4.5 displays export share of CCT, WE, EEL and SPVS for top five Asian countries in world export in 2002, 2005, 2008 and 2009. Table 4.5 has four (A, B, C and D) panels highlighting on major sub-categories of CFGT and provides corresponding top trade performers in Asia. Panel A in Table 4.5 displays the export share of CCT of top five Asian countries in world export in 2002, 2005, 2008 and 2009. Japan, Turkey and Singapore were leading exporters of clean coal technologies in top five Asian countries list in 2002, 2005, 2008 and 2009. Pakistan and India emerged as CCT exporter during global crisis period of 2008–2009. Japan held top rank in clean coal technologies in 2002 and 2005; however, Pakistan replaced Japan in 2008 and 2009. Singapore exported CCT consistently till the global crisis started (i.e., during 2002–2008). In Asia, CCT export share of Japan and Turkey declined during 2002–2008, while that of Singapore increased for the said period.

Panel B in Table 4.5 displays the export share of WE of top five Asian countries in world export in 2002, 2005, 2008 and 2009. Japan, China and Turkey were consistent exporters of wind energy equipment and products in Asia in our study periods. Japan held top position with nearly constant export share of WE in Asia till pre-global crisis. Turkey's export share of WE increased over time except in 2009. China's export share of WE declined in 2005; otherwise it more or less remained constant during 2002–2009.

Panel C in Table 4.5 displays the export share of EEL of top five Asian countries in world export in 2002, 2005, 2008 and 2009. From Asia, China was top in export share of energy-efficient lighting in world export in 2002, 2005, 2008 and 2009. China's export share of EEL increased from 0.0012% in 2002 to 0.00224% in 2009. China, Thailand, Sri Lanka and Macao were the leading EEL exporters during 2002–2009.

Panel D in Table 4.5 displays the export share of SPVS of top five Asian countries in world export in 2002, 2005, 2008 and 2009. From Asia, the top leading exporters of solar photovoltaic systems in world export during 2002–2009 were Japan, Malaysia, Thailand and Hong Kong. Japan was the top leader of SPVS in 2002 and 2005 while Malaysia replaced Japan in 2008 and 2009. Thailand's export share of SPVS in world export was nearly constant in all the study years. SPVS export share increased from 0.0038% in 2002 to 0.0053% in 2009.

Table 4.5 Export share of CCT, WE, EEL and SPVS in top five nations in Asian countries in world export in 2002, 2005, 2008 and 2009

Country	Export share in 2002	Country	Export share in 2005	Country	Export share in 2008	Country	Export share in 2009
A: CCT export share							
Japan	0.00086	Japan	0.00075	Pakistan	0.00092	Pakistan	0.00078
Turkey	0.00047	India	0.00062	Singapore	0.00076	Thailand	0.00025
Singapore	0.00020	Singapore	0.0003	India	0.00058	Turkey	0.00015
Russia	0.00017	Macao	0.00027	Japan	0.00057	Russia	0.00015
China	0.00009	Thailand	0.00021	Turkey	0.00016	Malaysia	0.00010
B: WE export share							
Japan	0.0028	Japan	0.00286	Japan	0.0027	China	0.00066
China	0.00064	Kyrgyz	0.00062	Turkey	0.00074	Turkey	0.00058
Singapore	0.00044	Turkey	0.00057	China	0.00061	Thailand	0.00023
Turkey	0.00038	Singapore	0.00044	South Korea	0.0006	Hong Kong	0.00021
South Korea	0.00027	China	0.00042	India	0.0005	Russia	0.00016
C: EEL export share							
China	0.0012	China	0.0013	China	0.002	China	0.00224
Thailand	0.00065	Sri Lanka	0.0009	Sri Lanka	0.00064	Thailand	0.00031
Sri Lanka	0.00017	Macao	0.00047	Macao	0.00042	Macao	0.00029
Japan	0.00013	Thailand	0.00037	Thailand	0.00033	Hong Kong	0.0002
South Korea	0.00012	Vietnam	0.00009	Hong Kong	0.00031	Turkey	0.00007
D: SPVS export share							
Malaysia	0.0115	Malaysia	0.0192	Japan	0.0114	China	0.0112
Japan	0.0091	Japan	0.0118	China	0.0107	Malaysia	0.0109
Thailand	0.0064	Thailand	0.0068	Malaysia	0.010	Hong Kong	0.0053
Hong Kong	0.0038	Hong Kong	0.0047	Macao	0.0067	Thailand	0.0046
Singapore	0.00256	China	0.0031	Hong Kong	0.0050	Macao	0.0039
China	0.00176	Singapore	0.0022	Thailand	0.0049	Philippines	0.0026

Source: Author's calculations

Table 4.6 shows the trends of export of sub-categories of CFGT in major regions in Asia during 2002–2009. Export shares of EEL and SPVS gradually increased in APTA region over the period of 2002–2009 (Table 4.6A). OC dominated in APTA and ASEAN regions for the said periods (Table 4.6A and Table 4.6B). CCT dominated in SAARC; however, export shares of CCT and SPVS improved over time in the period of 2002–2008 (Table 4.6C).

Table 4.6 Trends of export shares of sub-categories of CFGT in APTA, ASEAN and SAARC during 2002–2009

A: APTA

Year	CCT	WE	SPVS	EEL	OC
2002	0.007	0.051	0.16	0.083	1.83
2003	0.004	0.046	0.17	0.082	1.54
2004	0.005	0.038	0.21	0.085	1.64
2005	0.007	0.038	0.26	2.083	0.09
2006	0.010	0.040	0.35	0.112	1.81
2007	0.006	0.046	0.50	0.129	1.99
2008	0.008	0.059	0.84	0.144	2.11
2009	0.004	0.066	1.12	0.224	2.34

B: ASEAN

Year	CCT	WE	SPVS	EEL	OC
2002	0.012	0.024	0.64	0.019	1.211
2003	0.018	0.027	0.73	0.015	1.198
2004	0.024	0.025	0.75	0.013	1.175
2005	0.019	0.029	0.78	0.013	1.085
2006	0.022	0.026	0.65	0.011	1.111
2007	0.025	0.024	0.53	0.016	1.428
2008	0.034	0.026	0.53	0.009	1.228
2009	0.016	0.015	0.73	0.046	1.402

C: SAARC

Year	CCT	WE	SPVS	EEL	OC
2002	–	–	0.011	0.010	0.305
2003	0.012	0.011	0.067	0.007	0.538
2004	0.038	0.019	0.096	0.008	0.564
2005	0.049	0.020	0.089	0.010	0.707
2006	0.012	0.032	0.124	0.011	1.019
2007	0.028	0.035	0.164	0.019	1.106
2008	0.059	0.043	0.33	0.016	1.348
2009	0.073	0.002	0.001	0.0	0.149

Source: Author's calculations

4.3 Conclusion

Trade performance of CFGT in Asia increased rapidly, and an average growth rate of CFGT trade was around 8% during 2002–2017. Export shares of CFGT suggested that only few countries in Asia like Japan, Hong Kong and China had shares above the world average. Japan was the top exporter of WE and SPVS till 2008; however, China replaced Japan in 2009. China was at sixth position in SPVS exporter list in 2002; with raising export share of SPVS over time, China reached at second position in 2009. It is noted that East Asia was specialized in export of EEL and SPVS in the first decade of this century. Countries in Asia were CFGT importers from countries within and outside Asia. Import share of CFGT indicated that as many as ten developing economies of Asia had import shares above the world average. This provides indications of the realization of importing CFGT in Asia from the world.

Chapter 5
Comparative Advantage

Abstract This chapter identifies trade advantage on CFGT and its sub-categories in Asia and subregions. RCA and Michelaye indices indicate that China, Hong Kong and Japan had enjoyed comparative advantage in CFGT trade in the period of 2002–2017. Most of countries in Asia imported CFGT during 2002–2008; however, they did not enjoy comparative advantage in CFGT export for the said period.

Keywords RCA · CFGT · CCT · EEL · WE · SPVS · OC · SAARC · ASEAN · APTA · Emerging market · Pre-global financial crisis · Asia · China · Japan · Hong Kong · Indonesia · Thailand · Malaysia · India · Pakistan · Singapore · South Korea

Following the earlier chapter on performance of export and import in Asia, this chapter attempted to explore and identify trade advantage on CFGT in Asia and subregions. Trade performance basically depends on comparative advantage of a country for certain products like CFGT and its sub-categories.

5.1 Revealed Comparative Advantage in CFGT for Asia

Revealed comparative advantage (RCA) and Michelaye indices are used for comparative advantage. The study works out these indices which indicate comparative advantage of countries in Asia in CFGT trade. RCA is defined as the ratio of two shares[1] and $0 < RCA < +\infty$. A country is said to have a revealed comparative advantage if $RCA > 1$. Table 5.1 shows that RCA figures for Japan, China and Hong Kong had $RCA > 1$ in CFGT in 2008. Japan and Hong Kong had $RCA > 1$ in 2002. Again one might observed a rise of China from $RCA < 1$ in 2002 to $RCA > 1$ in 2008. China had a figure of 0.98 in 2002 and marginally improved to the figure of 1.31 in 2008.

---

[1]The numerator is the share of a country's total exports of the commodity of interest in its total exports. The denominator is the share of world exports of the same commodity in total world exports.

Table 5.1 Revealed comparative advantage of CFGT for top ten countries in Asia in 2002 and 2008

Country	2002	Country	2008
Japan	1.741	Japan	2.001
Hong Kong	1.111	China	1.313
China	0.986	Hong Kong	1.016
South Korea	0.893	South Korea	0.925
Singapore	0.718	Philippines	0.896
Malaysia	0.708	India	0.75
Thailand	0.690	Thailand	0.653
Turkey	0.467	Malaysia	0.612
Sri Lanka	0.258	Singapore	0.597
Russia	0.246	Macao	0.53

Source: Author's calculations

Table 5.2 RCA in EEL for top ten countries in Asia in 2002 and 2008

Country	2002	Country	2008
China	5.529	China	6.019
Thailand	2.990	Sri Lanka	1.922
Sri Lanka	0.796	Macao	1.264
Japan	0.593	Thailand	0.979
South Korea	0.558	Hong Kong	0.918
Hong Kong	0.311	India	0.480
Turkey	0.221	Vietnam	0.219
Bangladesh	0.205	South Korea	0.142
Macao	0.163	Japan	0.142
Russia	0.119	Turkey	0.126

Source: Author's calculations

5.1.1   Revealed Comparative Advantage in CFGT Sub-categories for Asian Nations in 2002 and 2008

Table 5.2 shows RCA > 1 for EEL for China, Sri Lanka and Macao in 2008 and RCA > 1 for China and Thailand in 2002. It indicates that the share of EEL exports in total exports of each of these countries was greater than the world share of EEL in the world total exports. RCA > 1 for China in 2008 are also reflected in the alternative Michelaye index for China with a positive figure. This reconfirms that China was performing better than other nations in Asia in EEL technologies. Similar thing happened with Macao in 2008.

Table 5.3 shows RCA in SPVS in 2008 and 2002 for top ten nations in Asia. Japan, China, Malaysia and Macao showed RCA > 1 in 2008, while Malaysia, Japan, Thailand, and Hong Kong had RCA > 1 in 2002. The figures show the rise of China and Macao in 2008 to levels reached in 2002.

Table 5.3 RCA in SPVS for top ten countries in Asia in 2002 and 2008

Countries	2002	Countries	2008
Malaysia	3.5719	Japan	2.2001
Japan	2.8355	China	2.0656
Thailand	1.9962	Malaysia	1.9253
Hong Kong	1.1939	Macao	1.2841
Singapore	0.7970	Hong Kong	0.9798
China	0.5457	Thailand	0.9356
South Korea	0.4271	Singapore	0.8044
Turkey	0.1950	India	0.7260
Russia	0.160	South Korea	0.5109
Sri Lanka	0.0572	Vietnam	0.3975

Source: Author's calculations

Table 5.4 RCA in CCT for top ten countries in Asia in 2002 and 2008

Countries	2002	Country	2008
Japan	0.8675	Pakistan	1.339
Turkey	0.4799	Singapore	1.117
Singapore	0.2028	India	0.8466
Russia	0.1769	Japan	0.8289
China	0.0912	Turkey	0.2339
Malaysia	0.0859	Russia	0.2106
Hong Kong	0.0383	Thailand	0.1573
Thailand	0.0281	Hong Kong	0.0686
South Korea	0.0270	Malaysia	0.0571
Sri Lanka	0.0011	South Korea	0.0475

Source: Author's calculations

Table 5.4 gives the figures for RCA for CCT of top ten countries in Asia in 2002 and 2008. It is noted that Pakistan and Singapore were the only countries in 2008 who had secured RCA > 1. India was at third rank with a value of 0.85. It seems that South Asian countries have developed expertise in CCT. It should be mentioned that no country in Asia had a comparative advantage in CCT in 2002.

Table 5.5 indicates that only Japan has a comparative advantage in the production of WE both in 2002 and 2008. Within Asia, Japan enjoyed comparative advantage in WE trade in the early twenty-first century.

Table 5.6 shows that Japan, the Philippines, China, Hong Kong and South Korea had comparative advantage in production of 'other codes' in 2008, while Japan and Hong Kong got values greater than one in 2002. None of the groups had RCA advantage in 2008 and 2002.

Table 5.7 provides RCA in CFGT export in selected major regional trade blocks in Asia in 2002 and 2008. Results of Table 5.7 suggest that RCA was greater than one in AFTA for EEL in 2002 and 2008 and RCA > 1 in 'other code' and SPVS were in favour of AFTA in 2008. RCA was in favour of ASEAN only in SPVS.

Table 5.5 RCA in WE for top ten countries in Asia in 2002 and 2008

Nations	2002	Nations	2008
Japan	2.580	Japan	2.043
China	0.592	Turkey	0.567
Singapore	0.402	China	0.468
Turkey	0.348	South Korea	0.462
South Korea	0.250	India	0.381
Russia	0.186	Singapore	0.326
Hong Kong	0.179	Thailand	0.219
Thailand	0.145	Hong Kong	0.154
Malaysia	0.034	Russia	0.111
Macao	0.008	Georgia	0.091

Source: Author's calculations

Table 5.6 RCA in other codes for top ten countries in Asia in 2002 and 2008

Nations	2002	Nations	2008
Japan	1.578	Japan	1.991
Hong Kong	1.156	Philippines	1.107
South Korea	0.993	China	1.078
China	0.962	Hong Kong	1.066
Singapore	0.714	South Korea	1.058
Thailand	0.669	India	0.752
Malaysia	0.494	Thailand	0.648
Turkey	0.475	Turkey	0.613
Sri Lanka	0.30	Singapore	0.588
Russia	0.268	Malaysia	0.569

Source: Author's calculations

5.1.2 Michelaye Index of CFGT for Selected Nations of Asia During 2002–2008

The Michelaye index is defined as the difference of two shares – the share of a country's total exports of the commodity of interest in its total exports and the share of the same country's imports of the same commodity in its total imports. A country is said to have a revealed comparative if the value of Michelaye index is positive. Table 5.8 displays the Michelaye index for selected countries for the period of 2002–2008, while Table 5.1 shows RCA for 2002 and 2008 only.

The Michelaye index has been worked out for some selected countries in Asia. It should be mentioned that countries, except Japan and Hong Kong, had negative values in almost all years from 2002 to 2007. This reinforces the point made above regarding comparative advantage in CFGT in Asia. However, nations in Asia might be importing regionally from some good performers (Hong Kong, Japan, the Philippines, Macao and China) or in few cases from outside Asia.

Table 5.7 RCA in CCT, WE, EEL and SPVS for regional trade blocks in 2002 and 2008

Regional groups	RCA in 2002	Regional groups	RCA in 2008
SPVS			
ASEAN	1.989	APTA	1.618
APTA	0.497	ASEAN	1.029
SAARC	0.034	SAARC	0.636
EEL			
APTA	3.805	APTA	4.295
ASEAN	0.892	SAARC	0.487
SAARC	0.480	ASEAN	0.283
Clean coal technologies			
ASEAN	0.123	SAARC	0.856
APTA	0.068	ASEAN	0.503
SAARC	0.0005	APTA	0.118
Wind energy			
APTA	0.468	APTA	0.457
ASEAN	0.220	SAARC	0.329
SAARC	0.0007	ASEAN	0.202
Other codes			
APTA	0.956	APTA	1.042
ASEAN	0.631	SAARC	0.666
SAARC	0.159	ASEAN	0.606

Source: Author's calculations

Table 5.8 Michelaye index of CFGT for some selected countries in Asia during 2002–2008

Year	JPN	HKG	MAC	IND	China	RUS
2002	0.022	0.002	−0.018		−0.013	−0.019
2003	0.024	0.001	−0.009	−0.006	−0.014	−0.020
2004	0.031	0.002	−0.009	−0.005	−0.016	−0.021
2005	0.030	0.002	−0.004	−0.005	−0.012	−0.0254
2006	0.029	0.002	−0.006	−0.0007	−0.009	−0.025
2007	0.032	0.003	−0.0017	−0.002	−0.007	−0.023
2008	0.035	0.001	0.006	0.0025	0.0007	−0.023

Source: Author's calculations

Michelaye Index of Sub-categories of CFGT for Selected Countries of Asia for the Period of 2002–2008

This section works out the Michelaye index for CFGT sub-categories for selected and identified countries of Asia. For convenience and comparison purpose, CFGT results are also reproduced. Table 5.9 shows the Michelaye index for CCT, SPVS, WE, EEL and OC for Japan. Most of the Michelaye indices for CCT, SPVS, WE and OC were positive in 2002, and all were remain positive except EEL during 2003–2008. It indicates strong comparative advantage of Japan in CCT, SPVS, WE

Table 5.9 Michelaye index for Japan, 2002–2008

Year	CFGT	CCT	SPVS	WE	EEL	OC
2002	0.022	0.00035	0.0069	0.0023	4.19E-05	0.016
2003	0.025	0.00017	0.0081	0.0021	−2.43E-06	0.017
2004	0.031	0.00031	0.0092	0.00207	−8.23E-05	0.024
2005	0.030	0.00033	0.0088	0.0023	−0.00011	0.022
2006	0.029	0.00018	0.0087	0.0020	−0.00012	0.021
2007	0.032	0.00014	0.0085	0.0018	−0.00016	0.024
2008	0.035	0.00016	0.0087	0.0020	−0.0002	0.026

Source: Author's calculations

Table 5.10 Michelaye index for China, 2002–2008

Year	CFGT	CCT	SPVS	WE	EEL	OC
2002	−0.0133	−2.96E-05	−0.00387	−0.0005	0.00108	−0.01276
2003	−0.01394	−9.93E-05	−0.00362	−0.0007	0.00118	−0.01285
2004	−0.01626	−0.000183	−0.00353	−0.0008	0.00125	−0.01507
2005	−0.01243	−0.000382	−0.00282	−0.001	0.00110	−0.01109
2006	−0.00933	−0.000178	−0.00175	−0.001	0.00137	−0.00936
2007	−0.00755	−0.00019	−6.8E-05	−0.0011	0.00164	−0.00945
2008	0.00078	−8.42E-05	0.00418	−0.0011	0.00198	−0.00556

Source: Author's calculations

Table 5.11 Michelaye index for Hong Kong, 2002–2008

Year	CFGT	CCT	SPVS	WE	EEL	OC
2002	0.00259	2.26E-06	−0.00108	−0.0001	−8.55E-05	0.00392
2003	0.00148	2.27E-07	−0.00125	−0.0001	−8.01E-05	0.00284
2004	0.00223	7.68E-06	−0.00110	-6E-05	−8.05E-05	0.00344
2005	0.00253	−4.03E-06	−0.00114	2.8E-06	−4.85E-05	0.00365
2006	0.0022	7.84E-06	−0.00127	6.3E-06	−5.87E-05	0.00352
2007	0.00289	2.41E-05	−0.00081	2.1E-05	−6.57E-06	0.00362
2008	0.00137	2.66E-05	−0.00116	5.5E-06	3.94E-07	0.00246

Source: Author's calculations.

and OC in Asia. Japan being an industrialized nation performs better in trade of CFGT which is basically a component trade to cleaner and energy-efficient technologies.

Table 5.10 shows that the Michelaye index is positive for EEL for China in the period of 2002–2008. So, China had a comparative advantage in EEL. Michelaye index was positive for China for CFGT and SPVS in 2008 only.

Table 5.11 indicates that Hong Kong performed better in CFGT and sub-categories like CCT and OC. Hong Kong improved in WE trade during 2005–2008.

Table 5.12 Michelaye index for Thailand, 2002–2008

Year	CFGT	CCT	SPVS	WE	EEL	OC
2002	−0.0166	−4.2E-05	0.0011	−0.0008	0.00058	−0.01716
2003	−0.0192	−0.00037	0.00227	−0.001	0.00046	−0.01917
2004	−0.0145	−0.00059	0.00146	−0.0009	0.00033	−0.0147
2005	−0.014	−0.00082	−0.00126	−0.0005	0.00030	−0.0145
2006	−0.0108	−0.00103	−7.981E-05	−0.0005	0.00019	−0.012
2007	−0.0044	−0.00076	−0.00143	−0.0005	0.00022	−0.0058
2008	−0.01069	−0.000224	−0.0016	−0.0005	0.00021	−0.012

Source: Author's calculations

Table 5.13 Michelaye index for Malaysia, 2002–2008

Year	CFGT	CCT	SPVS	WE	EEL	OC
2002	−0.01317	−0.0014	0.0043	−0.0004	−0.00016	−0.0142
2003	−0.01014	−0.00034	0.0068	−0.0006	−0.00019	−0.0126
2004	−0.00846	−0.000664	0.00969249	−0.0005	−0.00018	−0.0109
2005	−0.00848	−0.000953	0.0132	−0.0005	−0.00014	−0.0105
2006	−0.0087	−0.000681	0.0097	−0.0006	−0.00024	−0.0117
2007	−0.0094	−0.000814	0.00652	−0.0009	−0.00027	−0.0107
2008	−0.00707	−0.000671	0.00392	−0.0008	−0.00013	−0.0073

Source: Author's calculations

Table 5.14 Michelaye index for India, 2003–2008

Year	CFGT	CCT	SPVS	WE	EEL	OC
2003	−0.005894408	−0.000484	−0.0005131	−0.0012	−4.71E-05	−0.00444
2004	−0.004820726	0.000175	−0.0001512	−0.0009	−6.75E-05	−0.004432
2005	−0.005246931	0.000214	−8.782E-05	−0.0009	−9.71E-05	−0.004733
2006	−0.000792943	−9.5E-05	8.3266E-05	−0.0007	−9.51E-05	−0.000134
2007	−0.002248424	8.68E-05	5.0074E-05	−0.001	4.55E-05	−0.001636
2008	0.002514224	0.000311	0.00140527	−0.0012	5.62E-05	0.002117

Source: Author's calculations

Table 5.12 displays the Michelaye index for Thailand for the period of 2002–2008. The Michelaye index for Thailand shows that Thailand had comparative advantage in EEL in all years although it had a comparative disadvantage in CFGT during 2002–2008. Thailand enjoyed comparative advantage in SPVS for the period of 2002–2004.

Table 5.13 shows that figures for SPVS are positive for Malaysia, but like Thailand the figure indicates a comparative disadvantage in OC including CFGT.

Table 5.14 shows that Michelaye index was positive for CFGT, CCT, SPVS, EEL and for OC in 2008. The trade pattern of sub-categories of CFGT trade in India changed continuously since 2003. The Michelaye index for CCT was positive for all years except 2003 and 2006. Michelaye index for SPVS and EEL turned to positive onwards 2006 and 2007, respectively.

Table 5.15 shows the Michelaye index for Macao. Macao had positive figures for the CFGT, SPVS, WE and EEL for 2008.

Table 5.16 shows the Michelaye index for the Philippines, 2007–2008 for CFGT and its sub-categories. Michelaye index was positive for CFGT and OC; it was negative for WE and EEL.

Table 5.17 shows the Michelaye index for Vietnam. The values were positive for SPVS in 2007 and 2008 and EEL in all years except in 2006, while they were negative for all codes including CFGT.

Table 5.18 shows the comparative disadvantage of Sri Lanka in the production of CFGT and for its sub-categories except for EEL in 2007.

Table 5.19 shows that South Korea did not have comparative advantage in CFGT and its sub-categories. South Korea imported CFGT from other nations of APTA and the rest of the world.

Table 5.20 shows Michelaye index for Pakistan during 2002–2008. Pakistan did not have any comparative advantage in CFGT. Pakistan imported CFGT from other countries in SAARC and the world.

Table 5.21 below shows the Michelaye for Singapore. The values were positive for SPVS in 2007 and 2008.

Table 5.15 Michelaye index for Macao, 2002–2008

Year	CFGT	CCT	SPVS	WE	EEL	OC
2002	−0.017718833	0	−0.0041252	-1E-04	−0.000264	−0.013674
2003	−0.008954188	3.58E-05	−0.0016688	3.9E-05	−0.000121	−0.007348
2004	−0.008794925	−2.67E-07	−0.0014917	-4E-05	−0.00012	−0.007687
2005	−0.004110612	−3.09E-05	0.00026063	-3E-06	−4.32E-05	−0.004868
2006	−0.005859298	−0.002998	−0.0013226	-2E-05	0.000155	−0.003847
2007	−0.001722915	−3.09E-06	0.00274494	-4E-06	0.000144	−0.00545
2008	0.006184608	−5.1E-07	0.00474437	6.1E-05	0.000156	−0.000221

Source: Author's calculations

Table 5.16 Michelaye index for the Philippines, 2007–2008

Year	CFGT	CCT	SPVS	WE	EEL	OC
2007	0.009654928	−3.21E-05	0.00039559	−0.0002	−0.000176	0.0097314
2008	0.009417292	−7.21E-05	−8.778E-05	−0.0002	−0.000158	0.0099908

Source: Author's calculations

Table 5.17 Michelaye index for Vietnam, 2004–2008

Year	CFGT	CCT	SPVS	WE	EEL	OC
2004	−0.013252372	−0.001211	−0.0001244	−0.0008	2.8E-05	−0.011572
2005	−0.016520337	−0.001689	−0.0002469	−0.0009	4.61E-06	−0.014179
2006	−0.014582256	−0.002057	−0.0002536	−0.0008	−3.71E-06	−0.011986
2007	−0.027410192	−0.005977	0.00025249	−0.0009	0.000852	−0.02215
2008	−0.023887551	−0.001501	0.00016743	−0.001	3.48E-07	−0.022231

Source: Author's calculations

Table 5.18 Michelaye index for Sri Lanka, 2002–2008

Year	CFGT	CCT	SPVS	WE	EEL	OC
2002	−0.013487992	−1.19E-05	−0.0017256	−0.0002	−0.000579	−0.011645
2003	−0.005874594	−2.96E-05	−0.0012163	−0.0002	−0.000302	−0.004511
2004	−0.007036297	−6.91E-05	−0.0009699	−0.001	−9.36E-05	−0.006001
2005	−0.007213341	−1.44E-05	−0.000347	−0.0004	6.47E-05	−0.006272
2006	−0.009596398	−0.000101	−0.0010202	−0.0004	−0.000214	−0.007812
2007	−0.008345658	−0.002635	−0.0005292	−0.0004	0.00039	−0.004802
2008	−0.006069723	−1.57E-05	3.4152E-05	−0.0003	−0.000346	−0.004958

Source: Author's calculations

Table 5.19 Michelaye index for South Korea, 2002–2008

Year	CFGT	CCT	SPVS	WE	EEL	Other
2002	−0.009075749	−0.001043	−0.003856	−0.0009	−6.84E-05	−0.00504
2003	−0.018069444	−0.001522	−0.0045445	−0.0011	−9.85E-05	−0.012436
2004	−0.018367873	−0.00118	−0.004319	−0.0011	−0.000144	−0.013253
2005	−0.014877582	−0.000785	−0.0042082	−0.001	−0.000177	−0.010547
2006	−0.013289666	−0.000867	−0.0034181	−0.0011	−0.000182	−0.009144
2007	−0.021559932	−0.000572	−0.0041143	−0.001	−0.000166	−0.017658
2008	−0.01498956	−0.0005	−0.0045976	−0.0009	−0.000135	−0.010338

Source: Author's calculations

Table 5.20 Michelaye index for Pakistan, 2002–2008

Year	CFGT	CCT	SPVS	WE	EEL	Other
2003	−0.012334195	−0.000882	−0.0005089	−0.0004	−0.000186	−0.010688
2004	−0.011146644	−8.51E-05	−0.0003068	−0.0005	−0.000147	−0.009923
2005	−0.010033217	−0.00011	−0.0005023	−0.0005	−0.000117	−0.008977
2006	−0.012949758	−7.52E-05	−0.0005659	−0.0005	−7.49E-05	−0.011924
2007	−0.019968829	−0.002348	−0.0008803	−0.0007	−8.07E-05	−0.016411
2008	−0.025273825	−0.003625	−0.001125	−0.0008	−3.94E-05	−0.020258
2009	−0.034243714	−0.00366	−0.0016565	−0.0006	−0.000153	−0.029033

Source: Author's calculations

Table 5.21 Michelaye index for Singapore, 2002–2008

Year	CFGT	CCT	SPVS	WE	EEL	Other
2002	−0.01079	−0.00038	−0.00104	−7E-05	−8.97E-05	−0.0095
2003	−0.00726	−0.00077	−0.00040	−0.0002	−0.00012	−0.00586
2004	−0.0109	−0.00095	−0.00059	−0.0003	−0.00011	−0.0091
2005	−0.00603	−0.00025	−0.00033	−0.0003	−8.74E-05	−0.00499
2006	−0.00708	−2.56E-05	−0.00039	−0.0003	−6.34E-05	−0.00639
2007	−0.00715	−0.00034	5.36E-05	−0.0003	−8.7E-05	−0.00593
2008	−0.00503	−0.00027	0.00103	−0.0003	−6.91E-05	−0.00463

Source: Author's calculations

5.2 Conclusion

RCA and Michelaye index are worked out for comparative advantage analysis. Both indices suggest that Hong Kong, Japan and China enjoyed comparative advantage in CFGT; however, major countries in Asia did not have comparative advantage in CFGT export. Most of countries in Asia imported CFGT during 2002–2008.

Revealed comparative advantage for EEL was for China, Sri Lanka and Macao in 2008, while it was for China and Thailand in 2002. It suggests that the share of EEL exports in the total exports of each of these countries was greater than the world share of EEL exports. RCA > 1 for China in 2008 was also reflected in Michelaye index for China. This reconfirmed good performance of China in EEL. Japan, China, Malaysia and Macao showed greater than one RCA values in 2008 for SPVS, while Malaysia, Japan, Thailand, and Hong Kong had greater than one figures in 2002. Japan had a comparative advantage in the production of wind technology both in 2002 and 2008. Japan, the Philippines, China, Hong Kong and South Korea had a comparative advantage in production of 'other codes' in 2008, while Japan and Hong Kong got values greater than one in 2002.

Chapter 6
Competitiveness of CFGT

Abstract Competitiveness index has used to measure international market power in CFGT trade and its sub-categories in Asia. As per competitiveness index figures, Japan, China and Hong Kong were the most important economies from Asia in world export of CFGT in 2002 and 2008. Among the performers India, China and South Korea's competitiveness has improved their positions in the pre-global crisis period.

Keywords Competitiveness index · RCA · CFGT · CCT · EEL · WE · SPVS · OC · SAARC · ASEAN · APTA · Emerging market · Asia · China · Japan · Hong Kong · India · South Korea

6.1 Competitiveness

Competitiveness index (CI) is an indirect measure of international market power. CI is used here to judge the market power in CFGT trade and its sub-categories in Asia. CI is estimated as a ratio of individual country's export of CFGT to exports of CFGT by the world. Truly, competitiveness in trade is defined as the capacity of an industry to increase its share in international markets at the expenses of its rivals (UNESCAP Handbook, Trade Statistics in Policy Making 2007). CI is evaluated through a country's share of world markets in CFGT. $0 \leq CI \leq 100$, Higher value of CI indicates greater market power of the country. The figures and ranks in Table 6.1 show the competitiveness index of CFGT export for top ten economies in Asia in 2002 and 2008. The top five important economies in world export of CFGT from Asia in 2008 and 2002 were China, Japan, Hong Kong, South Korea and Singapore. Among countries in Asia, the competitiveness of China, India and South Korea improved in 2008 from 2002 position. China holds rank one in 2008 replacing Japan; similarly South Korea attained rank three replacing Hong Kong, while Singapore and Turkey remained at their respective ranks at fifth and ninth position. However, India was not in top ten in 2002, and India emerged at sixth rank in 2008.

Table 6.1 Competitiveness index for export of CFGT of top ten countries in Asia in 2002 and 2008

Country	Competitiveness Index 2002 (%)	Rank 2002	Country	Competitiveness Index 2008 (%)	Rank 2008
Japan	12.479	1	China	12.621	1
China	5.523	2	Japan	10.506	2
Hong Kong	3.859	3	South Korea	2.622	3
South Korea	2.496	4	Hong Kong	2.526	4
Singapore	1.546	5	Singapore	1.356	5
Malaysia	1.145	6	India	0.917	6
Thailand	0.809	7	Malaysia	0.817	7
Russia	0.451	8	Thailand	0.772	8
Turkey	0.287	9	Turkey	0.462	9
Sri Lanka	0.021	10	Russia	0.308	10

Source: Author's Calculations

6.2 CI for Trade in CFGT Sub-categories of Countries in Asia in 2002 and 2008

Competitiveness index shows the share of exports of one product by a country in world exports of the same product. CI shows the countries' international profile with respect to a product traded internationally. Table 6.2 shows the index for nations in Asia for EEL. China, Japan and Thailand were ranked at top three for 2002, while Hong Kong replaced Japan unchanging position of China and Thailand in 2008. It is notable the big gap between the figures of China and the second-ranked nation in 2002. This gap was widening in 2008. China's competitiveness was 57.84 in 2008, while, the second rank, Hong Kong's competitiveness was 2.28. CI of China was 30.96, while it was only 4.25 for Japan holding second rank in 2002.

Table 6.3 shows the competitiveness index in SPVS for countries in Asia in 2002 and 2008. Respectively, China, Japan and Malaysia ranked one, two and three in 2008, while Japan, Malaysia and Hong Kong hold the said ranking in 2002. As per CI of SPVS, major countries in Asia like China, Japan, Malaysia, Hong Kong, Singapore, South Korea and Thailand hold top seven positions in 2002 and 2008. It is clear that East and Southeast Asian nations enjoyed competitiveness in SPVS trade in the first decade of the twenty-first century.

Table 6.4 shows the competitiveness index for trade in CCT of countries in Asia. Japan, Singapore and India were at first, second and third rank in 2008, respectively. Top three ranking were hold by Japan, China and Singapore in 2002. Japan was ahead of all other nations in Asia in CCT trade in 2002 and 2008.

Table 6.5 gives the competitiveness index in trade in wind energy of nations in 2002 and 2008. Japan, China and Singapore were at rank one, two and three in 2002. Japan and China remained in their positions, and South Korea replaced Singapore at third rank in 2008. Japan, China, South Korea and Singapore were at

Table 6.2 Competitiveness index for trade in EEL of top ten countries in Asia in 2002 and 2008

Country	Competitiveness index 2002(%)	Ranking 2002	Country	Competitiveness index 2008(%)	Ranking 2008
China	30.97	1	China	57.843	1
Japan	4.25	2	Hong Kong	2.284	2
Thailand	3.50	3	Thailand	1.157	3
South Korea	1.56	4	Japan	0.746	4
Hong Kong	1.08	5	India	0.586	5
Russia	0.219	6	South Korea	0.403	6
Turkey	0.136	7	Russia	0.145	7
Sri Lanka	0.065	8	Malaysia	0.116	8
Malaysia	0.061	9	Turkey	0.112	9
Bangladesh	0.019	10	Sri Lanka	0.106	10

Source: Author's calculations

Table 6.3 Competitiveness index for trade in SPVS of top ten countries in Asia in 2002 and 2008

Country	Competitiveness index 2002(%)	Ranking 2002	Country	Competitiveness index 2008(%)	Ranking 2008
Japan	20.33	1	China	19.85	1
Malaysia	5.78	2	Japan	11.549	2
Hong Kong	4.15	3	Malaysia	2.572	3
China	3.06	4	Hong Kong	2.437	4
Thailand	2.34	5	Singapore	1.827	5
Singapore	1.71	6	South Korea	1.448	6
South Korea	1.19	7	Thailand	1.106	7
Russia	0.29	8	India	0.887	8
Turkey	0.12	9	Russia	0.27	9
Sri Lanka	0.005	10	Turkey	0.23	10

Source: Author's calculations

top four positions in 2002 and 2008 and interchanged their rank among them. Turkey and India hold fifth and sixth rank in competitiveness index for WE in Asia in 2008.

Table 6.6 provides the competitiveness index in trade in OC of countries in Asia in 2002 and 2008. Japan, China and Hong Kong were at rank one, two and three in 2008, while Japan, China and Hong Kong were at respective positions in 2002. Japan, China and Singapore were fixed up their position, while South Korea and Hong Kong interchanged their position in 2002 and 2008.

Table 6.4 Competitiveness index for trade in CCT of top ten countries in Asia in 2002 and 2008

Country	Competitiveness index 2002(%)	Ranking 2002	Country	Competitiveness index 2008(%)	Ranking 2008
Japan	6.22	1	Japan	4.35	1
China	0.51	2	Singapore	2.54	2
Singapore	0.44	3	India	1.034	3
Russia	0.3	4	Russia	0.66	4
Turkey	0.295	5	China	0.45	5
Malaysia	0.14	6	Turkey	0.21	6
Hong Kong	0.133	7	Thailand	0.19	7
South Korea	0.076	8	Pakistan	0.18	8
Thailand	0.033	9	Hong Kong	0.17	9
Sri Lanka	8.75E-05	10	South Korea	0.13	10

Source: Author's calculations

Table 6.5 CI for trade in wind energy of top ten nations in Asia in 2002 and 2008

Country	Competitiveness index 2002(%)	Ranking 2002	Country	Competitiveness index 2008(%)	Ranking 2008
Japan	18.293	1	Japan	10.726	1
China	3.279	2	China	4.496	2
Singapore	0.857	3	South Korea	1.309	3
South Korea	0.690	4	Singapore	0.740	4
Hong Kong	0.616	5	Turkey	0.503	5
Russia	0.337	6	India	0.466	6
Turkey	0.212	7	Hong Kong	0.382	7
Thailand	0.167	8	Russia	0.349	8
Malaysia	0.054	9	Thailand	0.259	9
Macao	0.0003	10	Malaysia	0.092	10

Source: Author's calculations

Table 6.7 shows the competitiveness index for export of CFGT and its sub-categories for major regional trade blocks in and around Asia in 2002 and 2008. Competitiveness for CFGT export in AFTA and SAARC improved in 2008 compared to 2002, while it was nearly unchanged in the ASEAN. Competitiveness of all sub-categories of CFGT trade increased in SAARC from 2002 to 2008. Maximum improvement in competitiveness is observed in AFTA. Competitiveness index of EEL increased from 32.61 in 2002 to 58.94 in 2008 in AFTA. Competitiveness of

Table 6.6 CI for trade in OC of top ten nations in Asia in 2002 and 2008

Country	Competitiveness index 2002(%)	Ranking 2002	Country	Competitiveness index 2008(%)	Ranking 2008
Japan	11.312	1	Japan	10.453	1
China	5.388	2	China	10.362	2
Hong Kong	4.017	3	South Korea	2.999	3
South Korea	2.774	4	Hong Kong	2.652	4
Singapore	1.538	5	Singapore	1.336	5
Malaysia	0.799	6	India	0.918	6
Thailand	0.784	7	Thailand	0.766	7
Russia	0.492	8	Malaysia	0.760	8
Turkey	0.292	9	Turkey	0.543	9
Sri Lanka	0.024	10	Philippines	0.365	10

Source: Author's calculations

Table 6.7 Competitiveness index for export of CFGT and its sub-categories by regional blocks in Asia and Asia-Pacific region in 2002 and 2008

Country	Competitiveness index 2002 (%)	Rank 2002	Country	Competitiveness index 2008 (%)	Rank 2008
CFGT					
APTA	8.043	1	APTA	16.174	1
ASEAN	3.500	2	ASEAN	3.388	2
SAARC	0.024	3	SAARC	0.945	3
EEL					
APTA	32.614	1	APTA	58.938	1
ASEAN	4.411	2	ASEAN	1.577	2
SAARC	0.084	3	SAARC	0.692	3
SPVS					
ASEAN	9.83	1	APTA	22.20	1
APTA	4.25	2	ASEAN	5.73	2
SAARC	0.006	3	SAARC	0.90	3
CCT					
ASEAN	0.608	1	ASEAN	2.8	1
APTA	0.586	2	APTA	1.62	2
SAARC	8.75E-05	3	SAARC	1.22	3
WE					
APTA	6.21E-05	1	APTA	6.272	1
ASEAN	1.69E-05	2	ASEAN	1.127	2
SAARC	1.77E-09	3	SAARC	0.468	3
OC					
APTA	8.190	1	APTA	14.295	1
ASEAN	3.120	2	ASEAN	3.377	2
SAARC	0.028	3	SAARC	0.946	3

Source: Author's Calculations

EEL also raised in SAARC, while it declined in ASEAN for the said period. Competitiveness index for SPVS was 4.25 in 2002 and improved to 22.2 in 2008 in AFTA, and in the ASEAN, it was 9.83 in 2002 and declined to 5.73 in 2008. However, in the ASEAN, competitiveness index of CCT, WE and OC increased marginally in 2008 compared to 2002. Competitiveness for CCT, WE and OC in AFTA also increased from 2002 to 2008. So, in regional blocks, AFTA and SAARC significantly improved the competitiveness index for export of CFGT and its sub-categories during 2002–2008.

6.3 Conclusion

The competitiveness index shows China, Hong Kong and Japan were the most important economies from Asia in world export of CFGT in 2002 and 2008. Among the performers India, China and South Korea's competitiveness has improved their positions during 2002–2008.

Part III
Regional Orientation

Chapter 7
Regional Orientation Index

Abstract This chapter finds out regional orientation index for regional trade blocks within Asia and indicates regional bias for CFGT having a value greater than one. It shows that most of countries in the ASEAN and APTA were importing from countries within Asia. The APTA was better in terms of EEL export. The ASEAN as a group has regional bias towards its own region for all codes except SPVS in 2002 and 2008. The APTA as a group has regional bias for OC and CCT in 2008. The SAARC as a group has regional bias for EEL in both years and CCT in 2008. India, Pakistan and Sri Lanka prefer to trade in CFGT regionally. Asia is biased for OC in 2002 and 2008 and CCT in 2008.

Keywords Regional orientation index · Michelaye index · Competitiveness index · RCA · CFGT · CCT · EEL · WE · SPVS · OC · SAARC · ASEAN · APTA · Emerging market · Asia · China · India · Japan · Pakistan · Sri Lanka · Singapore

7.1 Regional Orientation Index of CFGT Trade

Regional orientation index (ROI) is applied in regional trade blocks and forms certain policy opinion on countries' competitiveness, trade patterns, changing comparative advantage and regional bias. The study also finds that the trade of such CFGT has a regional bias for most of the countries in the region although almost all are net importers from Japan and Hong Kong and in 2004 onwards more from China.

The regional orientation index tells us whether exports of a particular product from the country under study to a given destination (its own regional grouping) are greater than exports of the same product to other destinations (the rest of the world). The index is the ratio of two shares.[1] $0 \leq ROI \leq + \propto$, i.e. ROI lies between 0 and $+ \infty$ (plus infinity). A value is greater than unity that suggests a regional bias in exports.

---

[1] The numerator is the share of a country's exports of a given product to the region of interest in total exports to the region. The denominator is the share of exports of the product to other countries in total exports to other countries (the rest of the world).

Table 7.1 Regional orientation index for CFGT for selected countries and regional groups in 2002 and 2008

	ASEAN		APTA		SAARC	
	2002	2008	2002	2008	2002	2008
China			0.73	1.28		
Malaysia	1.08	1.14				
Philippines	0.95***	0.62				
Singapore	2.12	1.81				
Thailand	1.13	1.37				
Vietnam	1.60**	1.15				
India			0.85*	0.39	1.16*	0.62
Pakistan					9.46*	1.02
Sri Lanka			2.25	2.26	1.37	1.88
South Korea			1.30	1.74		
ASEAN	1.51	1.42				
APTA			1.01	1.13		
SAARC					2.11	0.55

Source: Authors calculations. *2003, **2004, ***2007

This study finds out ROI for selected countries in Asia and its regional trade blocks such as the APTA, ASEAN and SAARC (see Table 7.1). The study finds regional bias considering its values greater than one. Table 7.1 suggests that there is a regional bias for CFGT for most of the countries in the ASEAN and APTA. It seems to suggest that countries in the ASEAN and APTA might not have comparative advantage and might be net importers, but most of them were importing from countries within the region, except few cases.

7.2 Michelaye Index of CFGT and Its Sub-categories for Regional Groups During 2002–2008

Michelaye index identifies the sectors in which an economy or a group has a comparative advantage. A country is said to have a revealed comparative advantage if the value exceeds zero. Michelaye index is worked out for the ASEAN, APTA and SAARC (see Tables 7.2, 7.3, and 7.4) for CFGT and its sub-categories. The positive figures for two codes indicate that SPVS and EEL are two sub-categories of CFGT in which Asia has comparative advantage. The earlier analysis has shown that most of nations in Asia did not have a comparative advantage in the production of CFGT; however, they were net importers of CFGT. The study identifies and works out the Michelaye index for those countries and regional groups whose values for sub-categories of CFGT work out to be positive. The results indicate that China, Hong Kong, India, Japan, Macao, Malaysia, the Philippines, Thailand and Vietnam performed better for some sub-categories in terms of export pattern to its own import

Table 7.2 Michelaye index for sub-categories of CFGT for the ASEAN during 2002–2008

Year	CCT	WE	SPVS	EEL	Other	CFGT
2002	−0.00061	−0.00038	0.001211	4.62781E-05	−0.01291	−0.01296
2003	−0.00055	−0.00052	0.002411	−5.39438E-06	−0.01122	−0.01111
2004	−0.0008	−0.00054	0.002785	−2.4677E-05	−0.01103	−0.01115
2005	−0.00068	−0.00044	0.003126	−7.10207E-06	−0.00952	−0.00943
2006	−0.00058	−0.00045	0.0024	−4.68238E-05	−0.00944	−0.00885
2007	−0.00102	−0.00054	0.001466	−4.74595E-06	−0.00727	−0.00757
2008	−0.00044	−0.00052	0.001178	−2.42636E-05	−0.00763	−0.00754

Source: Authors calculations

Table 7.3 Michelaye index for sub-categories of CFGT for APTA during 2002–2008

Year	CCT	WE	SPVS	EEL	Other	CFGT
2002	−0.00039	−0.00064	−0.00375	0.000675	−0.00992	−0.01162
2003	−0.00052	−0.00085	−0.00343	0.000699	−0.0114	−0.01367
2004	−0.0004	−0.00089	−0.00327	0.000744	−0.013	−0.01505
2005	−0.00042	−0.00098	−0.00267	0.000665	−0.00961	−0.01143
2006	−0.00033	−0.00095	−0.0017	0.000871	−0.00763	−0.00844
2007	−0.00028	−0.00103	−0.00065	0.001102	−0.00969	−0.00911
2008	−0.00016	−0.00109	0.00243	0.00135	−0.0048	−0.00115

Source: Author's calculations

Table 7.4 Michelaye index for CFGT sub-categories for SAARC during 2002–2008

Year	CCT	WE	SPVS	EEL	Other	CFGT
2002	−0.00083	−0.00022	−0.00099	−0.00031	−0.00943	−0.01163
2003	−0.0005	−0.00094	−0.00059	−9.6E-05	−0.00574	−0.00721
2004	9.99E-05	−0.00076	−0.0003	−9.9E-05	−0.00617	−0.00676
2005	0.000137	−0.0008	−0.00028	−0.00011	−0.00635	−0.007
2006	−0.00014	−0.00061	−0.00015	−0.00012	−0.00296	−0.00385
2007	−0.00035	−0.00091	−0.00024	1.01E-05	−0.00395	−0.00517
2008	−0.00015	−0.00112	0.001123	3.44E-05	−0.00046	−0.00068

Source: Authors calculations

pattern. For example, China, India, Japan, Macao, Thailand and Vietnam have positive Michelaye index for EEL during 2002–2008, while Japan, India, Macao and Malaysia performed better in SPVS.

Table 7.2 provides Michelaye index for the ASEAN. Only SPVS had positive figures for the period of 2002–2008. All other categories had negative figures indicating that the ASEAN as a group did better in terms of export pattern to its own import structure for the sub-category solar photovoltaic systems only.

The Michelaye index for the APTA is given in Table 7.3. Positive values come for only EEL. Thailand, Vietnam and Macao in the APTA performed better in terms of their export pattern during 2002–2008.

Michelaye index for SAARC is worked out, and figures are given below in Table 7.4. CFGT sub-categories clean coal technologies, solar photovoltaic systems and energy-efficient lighting showed some positive values for some years (2008) indicating that SAARC as a region was net importers of CFGT and sub-category goods from the rest of the world. The positive values indicate the changing trade pattern of these countries in SAARC towards producing and exporting cleaner technologies.

7.3   Regional Orientation of CFGT Sub-categories

Table 7.5 provides the regional orientation index for selected countries in Asia and some regional trade blocks for sub-categories of CFGT in 2002 and 2008. The value of regional orientation index exceeding unity implies a regional bias in exports. The results indicate that Malaysia had regional bias for export of CCT, WE and OC in the ASEAN in 2002 and 2008. The Philippines also had regional bias for export of WE and OC in 2007 and 2008 and that of EEL in 2008. Singapore had regional bias (in the ASEAN) for all codes in 2002 and 2008 except SPVS in 2008. Thailand had regional bias (in the ASEAN) for exports of CCT, WE and OC in 2002 and 2008. China had regional bias in the APTA for export of CCT in 2008 and EEL in 2002. The ASEAN as a regional trade block had regional bias towards its own region for all codes except SPVS in 2002 and 2008.

India had regional bias for EEL in SAARC for both years (2002 and 2008) and in CCT in 2008. Sri Lanka had a regional bias for export of SPVS and EEL in 2002 and 2008. SAARC as a regional trade block had regional bias for EEL in 2002 and 2008 and CCT in 2008.

India had regional bias for clean coal technologies in the APTA in 2008. South Korea had regional bias in APTA for OC and SPVS in 2002 and 2008. The APTA as a trade block had regional bias for OC and CCT in 2008.

7.4   Conclusions

The study finds out regional orientation index for regional trade blocks in and around Asia and indicates regional bias for CFGT having a value of more than one. It suggests that most of nations in the ASEAN and APTA were importing from countries within the region. It is interesting to find that some of the countries in SAARC like Pakistan, Sri Lanka and India (2002) would like to trade in CFGT regionally; however, no regional group had comparative advantage in the production of CCT in 2002. Pakistan and Singapore were the only countries in 2008 who had secured comparative advantage, and India was close at third position in CCT. SAARC countries had developed expertise in the production of CCT in 2008.

Table 7.5 Regional orientation index of export of sub-categories of CFGT for selected countries and regional groups in 2002 and 2008

Country(in regional grouping)	CCT 2002	CCT 2008	WE 2002	WE 2008	SPVS 2002	SPVS 2008	EEL 2002	EEL 2008	OC 2002	OC 2008
Malaysia(ASEAN)	1.86	6.15	6.64	7.17	0.32	0.28	0.14	0.13	1.58	1.45
Philippines(ASEAN)			1.44**	10.4	0.01**	0.003	0.27**	1.81	1.01**	0.64
Singapore(ASEAN)	3.50	20.1	1.36	2.63	2.22	0.72	5.14	5.47	2.08	1.88
Thailand(ASEAN)	54.56	2.66	9.37	2.06	0.89	0.88	1.27	0.91	1.17	1.63
China(APTA)	0.01	5.53	0.44	O.85	0.77	0.75	1.36	0.65	0.73	1.60
India(SAARC)	0.71*	6.64	0.37*	0.11	0.51*	0.56	11.74*	17.9	1.22*	0.47
Pakistan(SAARC)			0.09*	0.03		2.66			9.57*	1.66
Sri Lanka(SAARC)			6.83	0.0003	7.11	2.17	233.64	1.61	0.85	0.46
India(APTA)	0.07*	3.11	0.26*	0.94	0.33*	0.33	6.10*	0.39	0.91*	0.30
South Korea(APTA)	1.09	0.22	0.46	1.63	1.14	1.84	0.07	0.41	1.33	1.75
APTA(APTA)	0.23	2.85	0.38	1.01	0.91	0.63	0.84	0.44	1.07	1.39
ASEAN(ASEAN)	3.61	18.5	2.09	2.97	0.67	0.58	1.40	1.01	1.71	1.53
SAARC(SAARC)		4.41	12.01	0.09	9.08	0.52	78.35	1.36	1.35	0.44

Source: Author's calculations. *2003 figures, **2007 figures

Michelaye index is also calculated for the ASEAN, APTA and SAARC for CFGT and its sub-categories. Positive values are observed only for EEL. Performances of Thailand, Vietnam and Macao in APTA were better in terms of their export pattern. SPVS and CFGT have positive figures during 2002–2008 for ASEAN which indicate comparative advantage in sub-categories. All other categories have negative figures indicating that the ASEAN as a group does better in terms of export pattern to its own import structure for SPVS only. The ASEAN as a group has regional bias towards its own region for all codes except SPVS in 2002 and 2008. SAARC as a group has regional bias for EEL in both years and CCT in 2008. The APTA as a group has regional bias for OC and CCT in 2008. Asia is bias for OC in 2002 and 2008 and CCT in 2008.

The above analysis based on trade indices indicates the factual position of each country with respect to trade of CFGT and its sub-categories. Gravity analysis will help us answer the question as follows: Why do we see trends like the above? Do tariff, environmental projects and tied aid, economic size, endowments, policy, transparency, regulations or infrastructure matter, among others, matter more for trade of such clean technologies of Asia? For understanding the bilateral trade flows, we study the trade gravity model. The gravity analysis will be followed in the next chapters to explain determinants of CFGT exports and its sub-categories in Asia. Gravity model is used to explain the role of tariff barriers, preferential trading arrangements, economic size and endowments, environmental regulations, distance between trading partner and membership of multilateral agreement, among others, on trade of such CFGT and its sub-categories.

Part IV
Analysis

Chapter 8
Potential Business of Climate Friendly Goods and Technologies in Asia

Abstract This chapter examines empirical relationships, analyses determinants of CFGT trade in the precrisis period and predicts bilateral trade flows using the gravity model in Asia. Income level, geographical distance and developmental position of both trading partners and country characteristics, economic policy reforms and available infrastructure are important determinants of CFGT trade and its subcategories. The total estimated potential exports of CFGT within Asia and the European Union were around $32 billion US dollar (USD) and $10 billion US dollar in 2008, respectively. Estimating potential trade gap, this chapter predicts the value of trade opportunity of CFGT in Asia and identifies potential trading partners within and outside Asia.

Keywords CFGT · Clean coal technology · Energy-efficient technology · Wind energy · Solar photovoltaic system · Other code · SAARC · ASEAN · APTA · Gravity model · Potential trade opportunity · Export · Import · GDP · GDP per capita · Distance · Potential trade gap · Asia · China · Japan · India · South Asia

8.1 Introduction

Newly industrialized countries (NIC) in East Asian nations like Taiwan, South Korea and Hong Kong are experienced in trade-led growth in the last quarter of the twentieth century. Their exports were increased rapidly following imports of updated technologies from developed countries, which might reduce technological restrictions or limitations. They import new technologies through foreign collaborations and produce better quality of goods at comparatively low cost (due to available cheap labour) and finally export quality products embedded with upgraded technologies at competitive price. Trade model of East Asia is adopted by neighbouring and other Asian countries. Several Asian economies are emerging with trade diversity. Truly, import trade meets domestic demand as well as it also fulfils requirement for creation of export opportunities in emerging Asia in this twenty-first century. CFGT import might reduce technological restrictions of underdeveloped countries.

Availability and effective adaptation of the use of CFGT are essential to mitigate global climate change. CFGT export increased slowly in the period of 1996–2003; however, CFGT export rose rapidly after 2003, and it overtook CFGT import growth in Asia during 2004–2006. The share of CFGT export in total world export increased from 2.48% in 2002 to 2.71% in 2008 and slightly down to 2.68% in 2009, while world import of CFGT share rose from 2.4% in 2002 to 2.6% in 2008 (Dinda 2014a, b). CFGT trade share was low (around 7.5% of world merchandise export); however, it takes momentum after 2009.

Reporter country's export turns to be import of its partners. Asia's CFGT exports increase gradually with intraregional and interregional trade during 2002–2008. Intraregional demand was nearly 51% and only 49% for interregional demand of CFGT in 2008. It is true that internal demand within Asia is very high for CFGT, and it increases with economic development over time.

This chapter investigates stable empirical relationships (Learner and Levinson 1995) and estimate of bilateral trade flows applying the gravity model in Asia. The gravity model is used in this empirical analysis for determinants of the distribution of goods or production factors across space and economic size. Truly, the gravity model explains the role of economic size and resource endowments, distance between trading partners, membership of regional and multilateral agreements, among others on trade of such CFGT. In this chapter, this gravity model is used in several cross-sectional data analysis for estimating CFGT import and export in different times. Initially we examine the gravity equation considering the bilateral total trade of CFGT import for the year 2006 and later investigate CFGT exports for the year 2005 and finally analyse CFGT export and import in 2008. Economic growth momentum gained considerably in 2005–2006 and reached at maximum in 2008. The gravity model analysis is useful to explain determinants of import and export potential of CFGT for Asian countries within the region and interregional such as in the North America and the European Union (EU).

In our regression analysis, we have used the log values of all the variables except for dummies. In original version of Tinbergen (1962), the model is expressed in a log-log form. So the parameters are elasticity of the trade flow with respect to the explanatory variables. The least square econometric technique is used for the gravity Eq. (3.6) that is estimated for analysis purpose. Trade gap is measured the differences between estimated and actual bilateral trade flows. Untapped trade gap is identified as potential trade opportunity which may rise with reducing restrictions.

8.2 Empirical Findings and Analysis

Initially, we discuss on CFGT import in Asia in 2006. Model 1 is a basic CFGT import gravity model which consists of reporter country's GDP, partners' GDP, per capita GDP of reporting and partners, distance between pair countries and weighted tariffs. Country characteristics dummy variables are added to model 2. Policy, Infrastructure and FDI are incorporated in gravity model 3, 4 and 5, respectively.

Infrastructure and policy are the score Indies which are calculated on the basis of available information. Model 6 combines all variables.

Table 8.1 provides abovesaid six different estimated gravity models of import of CFGT in Asia in 2006. In model 1, coefficients of reporter country's GDP, GDP partner, geographical distance between two countries and constant term are statistically significant at 1% level. Import elasticity of CFGT in 2006 with respect to reporting country's GDP is 0.847 which is inelastic. It suggests that import of CFGT might increase by 0.847% if income of the reporting country increases by 1%. Import elasticity of CFGT with respect to the partner country's GDP is elastic (1.03), which suggests that if the partner country's GDP increases by 1%, import of CFGT increases by 1.03% (which is more than 1%).

Coefficient of partners' per capita GDP is significant at 5% level. Import elasticity of CFGT with respect to per capita GDP (development index) of partner country is inelastic (0.156). CFGT import increased by 0.156% as 1% per capita GDP increased in partner country. It is clear from these findings that import of CFGT increases with level of economic activities of both countries and development of partner. Coefficient of geographical distance between country pair is −0.814, which is negative as it is expected in the gravity model. Import reduces with increasing distance between trade partners. Coefficient of weighted tariffs is negative as expected; however, it is statistically insignificant,[1] and we left it without any comments. The constant term is statistically highly significant which suggests that certain explanatory variables are needed to explain the model. Considering only statistically significant coefficients of *base model 1*, the estimated CFGT import determinants in Asia in 2006 are

$$\ln M_{ij} = -29.467 + 0.847 \ \ln GDP_i + 1.03 \ \ln GDP_j \\ + 1.56 \ \ln pcgdp_j - 0.814 \ \ln DT_{ij} \tag{8.1}$$

Several dummy variables related to country's characteristics and regional agreements are added to the base model 1 for forming model 2, which represents a standard practice of the gravity model. In model 2, among additional variables (compared to model 1) coefficients of common official language and regional agreement are significant at 1% and 10% level, respectively. On the basis of statistically significant coefficients of *model 2*, the estimated CFGT import determinants in Asia in 2006 are

$$\ln M_{ij} = -32.2 + 0.911 \ \ln GDP_i + 0.999 \ \ln GDP_j \\ + 0.161 \ \ln pcgdp_i + 0.245 \ \ln pcgdp_j - 0.762 \ \ln DT_{ij} \tag{8.2} \\ + 1.16 D_{Com\_Office\_Lang} + 0.936 D_{Re\,gionalAgreement}$$

This study considers three major important variables such as infrastructure, policy and FDI, which are added in model 3, model 4 and model 6, respectively.

---

[1] It remains insignificant in all six models (see Table 8.1).

Table 8.1 Estimated gravity model for Import of CFGT in Asia in 2006

Variables	1	2	3	4	5	6	7
Constant	-29.467*** (-8.94)	-32.2*** (-9.93)	-37.382*** (-10.78)	-33.812*** (-10.6)	-32.114*** (-9.36)	-32.46*** (-7.45)	-32.734*** (-6.65)
lnGDP_Reporter	0.847*** (9.86)	0.911*** (10.85)	0.977*** (10.83)	1.004*** (11.45)	1.019*** (11.6)	0.819*** (4.7)	0.935*** (4.21)
lnGDP_Partner	1.03*** (13.61)	0.999*** (13.39)	1.096*** (14.35)	1.037*** (15.14)	0.983*** (13.27)	1.089*** (9.31)	1.101*** (10.14)
lnpcgdp_reporter	0.131 (1.54)	0.161** (1.97)	0.0949 (1.11)	0.031 (0.35)	0.038 (0.42)	0.137 (1.54)	0.041 (0.46)
lnpcgdp_partner	0.156** (1.97)	0.245*** (3.01)	0.024 (0.25)	-0.4287*** (-3.57)	-0.482*** (-3.94)	0.275*** (3.16)	-0.453*** (-3.66)
lnDistance	-0.814*** (-6.59)	-0.762*** (-6.1)	-0.892*** (-6.85)	-0.872*** (-7.14)	-0.869*** (-7.03)	-0.772*** (-6.09)	-0.858*** (-6.92)
lnTarifwt	-0.037 (-0.87)	-0.022 (-0.54)	0.005 (0.13)	0.019 (0.45)	0.0126 (0.31)	-0.002 (-0.03)	0.207 (0.43)
D Contiguous		0.204 (0.43)	0.193 (0.42)	0.215 (0.49)	0.291 (0.67)	0.221 (0.46)	0.339 (0.78)
DCommonofficial language		1.16*** (4.69)	0.704*** (2.67)	0.668*** (2.82)	0.81*** (3.25)	1.242*** (4.71)	0.955*** (3.59)
D Colony		-0.63 (-1.11)	-0.609 (-1.1)	-0.46 (-0.88)	-0.455 (-0.88)	-0.605 (-1.06)	-0.436 (-0.84)
D RegionalAgreement		0.936* (1.92)	0.643 (1.33)	0.442 (0.97)	0.409 (0.9)	0.852* (1.73)	0.340 (0.74)
lnPolicy_Score _Reporter			1.222** (2.26)		0.76 (1.12)	0.899 (1.2)	

lnPolicy_Score_Partner		1.959*** (3.83)		−1.358* (−1.88)	−1.471** (−2.03)
ln Infrastructure_score Reporter			1.34** (2.53)	0.726 (1.03)	0.388 (0.42)
ln Infrastructure_score Partner·			3.267*** (7.16)	4.245*** (6.21)	4.366*** (6.35)
ln FDI Reporter					0.126 (0.44)
					0.159 (0.65)
ln FDI Partner					−0.179 (−1.51)
					−0.128 (−0.99)
R²	0.5559	0.6221	0.6653	0.6714	0.6745
					0.5987
Adj R²	0.5461	0.6050	0.6502	0.6540	0.6546
					0.5806
Root MSE	1.5446	1.4409	1.356	1.3486	1.3475
					1.4848
N	279	279	279	279	279
					279

Note: Figures in parenthesis are t-values. '***', '**' and '*' denote the statistical level of significant at 1%, 5% and 10%, respectively

Policy score is measured on the basis of information available related to number of economic reform policy taking place and adopted in reporter and partner countries. Similarly infrastructure score is also calculated for both reporter and partner countries. Individually policy and infrastructure (of both report and partner countries) are positive and statistically highly significant in model 3 and model 4. Observing only statistically significant coefficients of *model 3* the estimated CFGT import determinants in Asia in 2006 is

$$
\begin{aligned}
\ln M_{ij} = &-37.382 + 0.977 \ \ln GDP_i + 1.096 \ \ln GDP_j - 0.892 \ \ln DT_{ij} \\
&+ 0.704 D_{Com\_Office\_Lang} + 1.222 \ \ln Policy Score_i \\
&+ 1.959 \ \ln Policy Score_j
\end{aligned}
\tag{8.3}
$$

Policy score is positive and highly responsive in both trading (reporter and partner) countries. Policy score is elastic in both reporting and partner nations. It also suggests that trading nations' economic policy reforms directly increase the import of CFGT in Asia. To capture the open economy market share, Asian nations build up infrastructure which has positive and direct impact on trade of both trading countries. Results indicate that infrastructure is highly elastic in both trading partners. Considering only statistically significant coefficients of *model 4*, the estimated CFGT import determinants in Asia in 2006 are

$$
\begin{aligned}
\ln M_{ij} = &-33.812 + 1.004 \ \ln GDP_i + 1.037 \ \ln GDP_j - 0.4287 \ \ln pcgdp_j \\
&- 0.872 \ \ln DT_{ij} + 0.668 D_{Com\_Office\_Lang} + 1.34 \ \ln Infrastructure Score_i \\
&+ 3.267 \ \ln Infrastructure Score_j
\end{aligned}
\tag{8.4}
$$

In model 5, coefficient of partner's infrastructure is positive and highly significant at 1% level, while policy is significant at 10% level with a negative coefficient. It is noted that infrastructure of partner nation is significantly positive; it suggests that import of CFGT in Asia directly depends on partner's infrastructure. Findings of model 6 and 7 suggest that FDI in Asia has no role to explain CFGT import in 2006. Coefficient of partner's policy score is negative and highly significant at 5% level; however, coefficient of infrastructure of partner is positive and highly significant in model 7. As per model fitting criteria, both R^2 and adjusted R^2 of model 7 are higher than other models. Root mean square error (RMSE) is the least in model 5 and very close to model 7. However, model 7 is the best fitted model considering R^2 and adjusted R^2 and RMSE. Considering only statistically significant coefficients of *model 7*, the estimated CFGT import determinants in Asia in 2006 are

$$
\begin{aligned}
\ln M_{ij} = &-32.73 + 0.93 \ \ln GDP_i + 1.1 \ \ln GDP_j - 0.45 \ \ln pcgdp_j - 0.86 \ \ln DT_{ij} \\
&+ 0.95 D_{Com\_Office\_Lang} - 1.47 \ln Policy_j + 4.37 \ln Infrastructure_j
\end{aligned}
\tag{8.5}
$$

Partners' countries economic position, infrastructure and policy reforms are major determinants of overall CFGT import in Asia in 2006. Do these determinants vary or remain the same for sub-categories of CFGT in Asia in 2006? Now we investigate determinants of import of major four (SPVS, CCT, EEL and WE) sub-categories of CFGT in Asia in 2006. Table 8.2 shows the estimated results of the gravity equation for import of CFGT sub-categories such as SPVS, CCT, EEL and WE in Asia in 2006. From Table 8.2 we find that the coefficients of GDP of partner and reporting countries, partner's per capita GDP, distance, common official language, policy of both countries, partner's infrastructure, and FDI inflow to reporting country are significant determinants of import of SPVS in Asia in 2006. Considering only significant coefficients, the estimated SPVS import determinants in Asia in 2006 are

$$
\begin{aligned}
\ln M_{ij} = &-38.03 + 0.645 \ln GDP_i + 1.181 \ln GDP_j - 0.666 \ln pcgdp_j - 1.19 \ln DT_{ij} \\
&+ 1.31 D_{Com\_Office\_Lang} + 2.86 \ln Policy_i - 1.882 \ln Policy_j \\
&+ 5.794 \ln Infrastructure_j + 1.07 \ln FDI_i
\end{aligned} \tag{8.6}
$$

Significant determinants of CCT import in Asia are GDP of partner country, per capita GDP of reporter and partner, distance, partner's policy and infrastructure and FDI inflow to reporter country. Considering only significant coefficients, the estimated CCT import determinants in Asia in 2006 are

$$
\begin{aligned}
\ln M_{ij} = &-16.8 + 0.834 \ln GDP_j - 0.227 \ln pcgdp_i - 0.307 \ln pcgdp_j \\
&- 0.866 \ln DT_{ij} - 1.597 \ln Policy_j \\
&+ 3.367 \ln Infrastructure_j + 0.864 \ln FDI_i
\end{aligned} \tag{8.7}
$$

Similarly, the estimated import of EEL determinants in Asia in 2006 is

$$
\begin{aligned}
\ln M_{ij} = &-34.56 + 1.074 \ln GDP_i + 1.133 \ln GDP_j - 0.321 \ln pcgdp_j \\
&- 1.248 \ln DT_{ij} + 0.535 D_{Com\_Office\_Lang} \\
&+ 2.291 \ln Infrastructure_i + 3.044 \ln Infrastructure_j
\end{aligned} \tag{8.8}
$$

and the estimated wind energy import determinants in Asia in 2006 are

$$
\begin{aligned}
\ln M_{ij} = &-33.4185 + 0.8384 \ln GDP_i + 1.151 \ln GDP_j - 0.757 \ln DT_{ij} \\
&+ 0.769 D_{Com\_Office\_Lang} + 2.976 \ln Infrastructure_j
\end{aligned} \tag{8.9}
$$

Determinants of EEL import in Asia are GDP of partner and reporter countries, partner's per capita GDP, distance, common official language and infrastructure of reporter and partner, while WE import is determined by GDP of partner and reporter countries, distance, common official language and infrastructure of partner. Partner's GDP is a common significant determinant for all sub-categories of CFGT import. Coefficient of GDP of reporter country is statistically highly significant for import of EEL and WE, while it is significant at low (10%) level in SPVS import and insignificant in case of CCT import, i.e. CCT import does not depend on importing country's income level. Income of reporting country is an important determinant for import of energy-efficient lighting and wind energy in Asia in 2006.

Coefficient of geographical distance between pair countries is negative and statistically significant in four major sub-categories (SPVS, CCT, EEL and WE) as per expected in the gravity model. Distance is highly sensitive (or elastic) in import of

Table 8.2 Estimated gravity model for the import of sub-categories of CFGT like SPVS, CCT, EEL and WE in Asia in 2006

Variables	Solar photovoltaic system (SPVS)	Clean coal technology (CCT)	Energy-efficient lighting (EEL)	Wind energy (WE)
Constant	**−38.03*****	**−16.8****	**−34.56*****	**−33.4185*****
	(−5.11)	**(−2.43)**	**(−6.7)**	**(−6.14)**
lnGDP_reporter	**0.645***	0.38	**1.074*****	**0.8384*****
	(1.94)	(1.22)	**(4.56)**	**(3.41)**
lnGDP_partner	**1.181*****	**0.834*****	**1.133*****	**1.151*****
	(7.08)	**(5.36)**	**(12.26)**	**(10.16)**
lnpcgdp_reporter	0.17	**−0.227****	−0.038	−0.01588
	(1.17)	**(−1.97)**	(−0.29)	(−0.16)
lnpcgdp_partner	**−0.666*****	**−0.307***	**−0.321****	−0.17919
	(−3.49)	**(−1.78)**	**(−2.53)**	(−1.35)
lnTarifwt	0.018	−0.06	0.0698	0.004676
	(0.26)	(0.88)	(1.3)	(0.09)
lnDistance	**−1.19*****	**−0.866*****	**−1.248*****	**−0.757*****
	(−6.26)	**(−3.84)**	**(−7.18)**	**(−5.52)**
D contiguous	0.297	0.576	−0.094	−0.093
	(0.44)	(1.21)	(−0.19)	(−0.2)
D_CommonOfficial Language	**1.31*****	0.163	**0.535****	**0.769*****
	(3.2)	(0.48)	**(2.06)**	**(2.69)**
D_Colony	−0.437	−0.42	−0.174	−0.45
	(−0.55)	(−0.87)	(−0.35)	(−0.82)
D_Regional Agreement	0.844	−0.211	0.119	0.436
	(1.2)	(−0.31)	(0.23)	(0.83)
lnPolicy reporter	**2.86****	0.473	0.779	−0.2415
	(2.48)	(0.5)	(1.0)	(−0.3)
lnPolicy partner	**−1.882***	**−1.597***	−0.834	−0.833
	(−1.69)	**(−1.66)**	(−1.1)	(−1.05)
Ln infra reporter	−1.896	1.219	**2.291****	1.36
	(−1.32)	(0.99)	**(2.11)**	(1.37)
Ln infra partner	**5.794*****	**3.367*****	**3.044*****	**2.976*****
	(5.49)	**(3.57)**	**(4.2)**	**(3.92)**
Ln FDI reporter	**1.07****	**0.864****	0.031	0.1604
	(2.44)	**(2.15)**	(0.10)	(0.5)
Ln FDI partner	−0.037	0.024	−0.193	−0.158
	(−0.20)	(0.15)	(−1.61)	(−1.24)
N	279	128	172	259
R²	0.6044	0.6549	0.7613	0.6316
Adj R²	0.5803	0.6052	0.7366	0.6073
Root MSE	2.0707	1.1837	1.0673	1.4229

Note: Figures in parenthesis are t-values. '***', '**' and '*' denote the statistical level of significant at 1%, 5% and 10%, respectively

SPVS and EEL, while it is less sensitive (or inelastic) in case of import of CCT and wind energy. Common official language is significant in all sub-categories except CCT. Common official language is a good indicator for easy communication between pair of trading countries. Coefficient of reporter's policy is statistically significant only for SPVS imports, while partner's policy is significant for SPVS and CCT imports. SPVS import depends on both traders' policy reforms. Coefficient of FDI inflow to reporter country was statistically significant only for import of SPVS and CCT. So, FDI inflow played an important role for importing SPVS and CCT in Asia. Coefficient of reporting country's infrastructure is statistically significant only for import of EEL, while coefficient of partner's infrastructure is for SPVS, CCT, EEL and WE. So, all major sub-categories of CFGT imports depend on infrastructure of trading partners. Overall imports of SPVS, CCT, EEL and WE depend on traders' income level, partner's economic development and infrastructure. However, imports of SPVS and CCT depend directly on FDI inflow in Asia. So, SPVS and CCT entered in Asia in 2006 through FDI channel.

Tables 8.1 and 8.2 suggest that imports of CFGT and its sub-categories in Asia are determined by income of both reporter and partner countries, economic development of partners associated with their policy reforms and infrastructure and common official language. So, import of CFGT in Asia crucially depends on economic positions of trading partners, infrastructure setup, policy reforms and common communicating language.

Imports generally boost up exports in emerging and developing economies in the follow-up periods. Now we investigate CFGT export determining factors in pre- and post-CFGT import in Asia in 2006. For the said purpose, we examine CFGT export in 2005 and 2008.

Table 8.3 presents the estimated results of the gravity equation for CFGT export in Asia in 2005 and 2008. Column 2–4 and column 5–7 of Table 8.3 provide results of CFGT export in 2005 and 2008, respectively. Row-wise Table 8.3 has three parts displaying estimated gravity equation of CFGT export in Asia in 2005 and 2008, their ANOVA in middle part and regression statistics at bottom part. We discuss first the fitting criteria, analysis of variance (ANOVA), and lastly estimated results. Overall fitting of the gravity equation is good in the cross-sectional data analysis (multiple R is 0.68257 in 2005 and 0.67924 in 2008; for more details, see bottom part of Table 8.3). R^2 is a fitting criterion that provides strength of association between actual and estimated dependent variables. In 2005, R^2 value of 0.4659 means that only 46.59% of the variations in CFGTs export is explained by the variables used in the equation, while R^2 value of 0.4745 suggests that variables used in the equation explained only 47.45% of the variations in CFGT exports in Asia in 2008. Adjusted R^2 (after adjustment with DF) is 0.4631 in 2005, while it is 0.4708 in 2008. Both F statistics (164.53 in 2005 and 128.97 in 2008) in ANOVA are statistically highly significant. Table 8.3 shows point estimation of coefficients with their corresponding statistical significance level marked with stars (as significance levels at 1%, 5% and 10%).

Table 8.3 Estimated gravity equation of CFGT export of Asia in 2005 and 2008

Variables	Export 2005			Export 2008		
	Coefficients	Standard error	t stat	Coefficients	Standard error	t stat
Intercept	**−43.24***	1.5323	−28.22	**−48.688***	1.765	−27.78
lnGDP_reporter	**1.5267***	0.0419	36.46	**1.407***	0.0471	24.86
lnGDP_partner	**0.8825***	0.0336	26.27	**0.904***	0.0366	24.68
lnpcgdp_reporter	**−0.195***	0.0467	−4.18	0.097	0.060	1.62
lnpcgdp_partner	−0.0620	0.047	−1.32	**−0.188***	0.0528	−3.56
Lndistw	**−1.2852***	0.0985	−13.04	**−0.538***	0.1077	−5.00
Contiguity	**0.7472***	0.3931	1.90	**1.007****	0.419	2.40
comlang_office	0.3459	0.3423	1.01	0.334	0.535	0.62
comlang_ethno	0.3117	0.304	1.025	0.242	0.501	0.48
Colony	0.4533	0.7223	0.63	**1.458***	0.756	1.93
Common colony	0.2170	0.228	0.95	**1.362***	0.2465	5.52
col45	1.0892	0.8791	1.24	0.283	0.9176	0.31
Smctry	**1.5052****	0.7361	2.045	0.7768	0.91	0.85
ANOVA						
	Sum of square	Mean sum of square	F stat	Sum of square	Mean sum of square	F stat
Regression	19629.26	1509.943	164.531	11874.475	989.54	128.97
Residual	22502.62	9.177253		13150.923	7.6726	
Total	42131.88			25025.398		
Regression statistics						
Multiple R	0.6826			0.6792		
R²	0.4659			0.4745		
Adjusted R²	0.4631			0.4708		
Standard error	3.0294			2.77		
Observations	2466			1727		

Note: Figures in parenthesis are t-values. '***', '**' and '*' denote the statistical level of significant at 1%, 5% and 10%, respectively

In 2005, the coefficients of reporter country's GDP, GDP partner, per capita GDP of reporter, geographical distance between two countries and constant term are statistically significant at 1% level. The coefficient of dummy for small country group is significant at 5% level, and coefficient of dummy for country group of contiguity is significant at a 10% level. Considering only statistically significant coefficients, the estimated CFGT export determinants in Asia in 2005 are.

$$\ln X_{ij} = -43.24 + 1.5267 \ \ln GDP_i + 0.8825 \ \ln GDP_j - 0.195 \ \ln pcgdp_i$$
$$-1.2852 \ \ln DT_{ij} + 0.7472 D_{contiguity} + 1.5052 D_{smctry} \tag{8.10}$$

CFGT export elasticity with respect to GDP of the reporting country in 2005 is elastic which suggests that export of CFGT would increase by more than 1.5% if income of the reporting country increases by 1%. CFGT export elasticity with respect to the partner country's GDP is inelastic (0.88), which suggests that if the partner country's GDP increases by 1%, the export of CFGT increases by 0.88% (<1%) in the reporter country's GDP. From this, one can guess that one part of partner country's internal demand is fulfilled by their production of CFGT. CFGT export elasticity with respect to per capita GDP (development index) of the reporting country is inelastic (−0.195). CFGT export decreases by 0.195% as 1% per capita GDP increases in reporting country. It is clear from these findings that export of CFGT increases with GDP while declines with per capita GDP (proxy of economic development). It is possibly due to the increase in internal demand of CFGT due to the raising awareness of global climate change and related policies and further provides the opportunity to produce CFGT in Asia. It indicates that opportunity of green business in Asia grows in 2005, and business of CFGT expands.

The coefficient of distance between country pair is negative as it is expected in the gravity model. Here, CFGT export elasticity with respect to distance was elastic (i.e. estimated coefficient of distance variable is −1.285) in 2005 and highly sensitive with distance.[2] The estimated coefficient of contiguity dummy variable is 0.747. CFGT exports are likely to be more in contiguous countries than others. Overall, CFGT exports are statistically significant in small countries in Asia in 2005. The constant term is statistically highly significant.

In 2008, the coefficients of GDP of reporter and partner, coefficient of per capita GDP of partner, distance between two countries and common colony are statistically significant at 1% level. The coefficient of dummy variable for contiguity and colony are statistically significant at a 5% and 10% level, respectively.

Considering only statistically significant coefficients, the estimated export of CFGT determinants in Asia in 2008 is

$$\ln X_{ij} = -48.688 + 1.407 \ \ln GDP_i + 0.904 \ \ln GDP_j$$
$$-0.188 \ \ln pcgdp_j - 0.538 \ \ln DT_{ij}$$
$$+1.007 D_{Contiguity} + 1.362 D_{Comcol} + 1.458 D_{Col} \tag{8.11}$$

CFGT export elasticity in 2008 with respect to GDP of reporting country is elastic (1.407) which suggests that CFGT export would increase by more than 1.4% if income of the reporting country increases by 1%. CFGT export elasticity with respect to the partner country's GDP is inelastic (0.904), which suggests that the reporter country's CFGT export increases by 0.904% if the partner country's GDP

---

[2]An increase in bilateral trade is explained as transportation cost decreases.

increases by 1%, in 2008. From these findings, one can guess that one part of part-ner country's internal demand is fulfilled by their CFGT production. CFGT export elasticity with respect to per capita GDP (development) of partner country is nega-tive and inelastic (-0.188). CFGT export decreases by 0.188% as 1% per capita GDP increases in partner country in 2008. It is clear from these findings that CFGT export increases with GDP while declines with per capita GDP or economic devel-opment. It is possibly due to the increase in internal demand of CFGT due to the raising awareness of global climate change and related policies and further provides the opportunity to produce CFGT in Asia. It indicates that opportunity of green business in Asia grows and expands CFGT business in 2008 in Asia.

The coefficient of distance between reporter and partner countries is negative and highly significant. Here, CFGT export with respect to distance[3] is inelastic (i.e. -0.538). There is a negative association between geographical distance and trade, i.e. bilateral trade rises with reducing transportation cost. CFGT exports are more among contiguity, common colonies and colony countries compared to others; it may be due to probably common administrative system and similar infrastructure in common colonial countries. It should be noted that estimated several coefficients of CFGT export in 2008 are different from that of in 2005. Country characteristic vari-ables like colony and common colony are significant in 2008 where as these are insignificant in 2005. Contiguity is highly significant in 2008 and significant at low level in 2005. Small country dummy is significant in 2005; however, it is insignifi-cant in 2008. The magnitude of coefficient of distance reduces from -1.285 in 2005 to -0.538 in 2008. This suggests that probably the cost of CFGT trade declines in 2008 compared to 2005. Coefficient of per capita GDP of reporter is significantly negative in 2005 and that of partner in 2008. Constant term is highly statistically significant which might not capture other unknown factors.

Considering per capita GDP as development index, results of Table 8.3 suggest that CFGT export reduced in 2005 with reporting country's development, while it declined in 2008 with development of partner country. It indicates that reporting country might absorb more CFGT and reduced its export in 2005; however, it was completely an opposite picture in 2008. With partner's development reporting coun-try's export declined in CFGT which is directly connected with import of trading partners in 2008. So, we have to examine import determinants of CFGT in 2008. Once again we examine import determinants of CFGT with the parity of export determinants in 2008. Table 8.4 provides the estimated results of gravity equation of CFGT import in Asia in 2008.

Considering only statistically significant and estimated CFGT import gravity model for Asia in 2008 is

---

[3] Literature (Disdier and Head 2008, Balassa 1966, Balassa and Bauwens 1987) supports these observations.

Table 8.4 Estimated results of the gravity model of CFGT import in Asia in 2008

Variables	Coefficients	Standard Error	t Stat
Intercept	−36.57***	1.819	−20.1
lnGDP_reporter	0.542***	0.047	11.48
lnGDP_partner	1.226***	0.041	29.83
lnpcgdp_reporter	0.354***	0.059	6.03
lnpcgdp_partner	0.666***	0.058	11.42
Lndistw	−1.416***	0.111	−12.74
Contiguous	0.924**	0.405	2.28
comlang_office	1.508***	0.499	3.02
comlang_ethno	−0.324	0.47	−0.69
Colony	−1.863***	0.658	−2.83
Comcol	0.245	0.289	0.85
Curcol	−7.052**	2.821	−2.5
col45	2.685***	0.821	3.27
Smctry	0.054	0.776	0.07
R^2	0.6022		
Adjusted R^2	0.5984		
RMSE	2.6341		
Observations	1367		

Note: '***', '**' and '*' denote the statistical significant level at 1%, 5% and 10%, respectively

$$\ln M_{ij} = -36.57 + 0.542 \ \ln GDP_i + 1.226 \ \ln GDP_j + 0.354 \ln pcgdp_i$$
$$+ 0.666 \ \ln pcgdp_j - 1.416 \ \ln DT_{ij} + 0.924 D_{Contiguous}$$
$$+ 1.508 D_{ColOfficial\ Lang} - 1.86 D_{Col} + 2.685 D_{Col45} - 7.052 D_{CurCol} \quad (8.12)$$

Overall, determinants of CFGT import in Asia in 2008 are directly related to their income levels (reporter's GDP and partner's GDP), development positions (reporter's per capita GDP and partner's per capita GDP), contiguity, common official language, colony45 and inversely related to colony and current colony. Comparing results of CFGT import in Table 8.4 and CFGT export in (right part of) Table 8.3, we observe that CFGT import determinants are different[4] from that of CFGT export in Asia in 2008. These are trade determinants in Asia just before the global financial crisis. Are the trade determinants changed in the crisis period? In this context we also investigate the trade determinants in Asia in 2009.

In the *global financial crisis*, in 2009, the coefficients of reporter's GDP, partner's GDP, distance among pair countries, colony, common colony and common official language are significant at 1% level, while that of contiguous is significant at 10% level. Considering only statistically significant coefficients, the estimated export of CFGT determinants in Asia in 2009 is

---

[4] Export and import trading partners could be different.

$$\ln X_{ij} = -44.57 + 1.44 \ \ln GDP_i + 0.82 \ \ln GDP_j - 0.97 \ \ln DT_{ij}$$
$$+ 0.67 D_{Contig} + 1.48 D_{Col} + 1.06 D_{Comcol} + 0.86 D_{ColOffLang} \quad (8.13)$$

CFGT export elasticity with respect to GDP of reporting country in 2009 is 1.44 which is elastic, while it is inelastic (0.82) with respect to partner's GDP. Country features are significant determinants of CFGT exports in 2009. More or less major determinants are remained same, however, magnitudes change.

8.2.1   Potential Trade Gap

Using the estimated export gravity Eq. (8.2), we predict the estimated CFGT export value of the reporting country with its trade partners in 2008. In this context, we define potential CFGT export gap as difference between the actual and predicted export value. *Potential trade gap in CFGT* indicates possible scope of raising CFGT trade with its partner (see Dinda 2014a, b). For example, in 2008, the estimated CFGT export in Asia was nearly $32.6 billion US dollar (USD); however, the actual CFGT export was around $23.4 billion USD; hence, the export gap was approximately $9.2 billion USD in 2008 (it is different from Dinda 2014a, b). So, trade opportunity value of CFGT export was around $9.2 billion USD in Asia in 2008. It indicates under performance of CFGT export of several Asian nations in 2008. This trade gap also suggests that those underperforming countries could raise their CFGT export value around $9.2 billion USD with their existing trade partners in 2008. In other word, potential trade opportunity was nearly $9.2 billion USD in CFGT export in Asia in 2008. India was on top having potential untapped CFGT export of around $5 billion USD in 2008, and other countries were followed by Russia, Pakistan, Hong Kong, etc. These major countries have huge untapped potential trade of CFGT. Intra- and inter-region groupings are done according to the partner country belonging to Asia, the EU, America, etc., and it identifies individual trade partners of the reporting country.

Intraregional demand for CFGT was also very high. Asia was net CFGT importer during 2002–2008 that reflected high demand for CFGT. Actual CFGT import within Asia was around $61 billion USD in 2008, and the potential CFGT import gap within Asia was approximately $20 billion USD, which was higher than CFGT export gap (see World Bank 2008, Dinda 2014a, b). Within Asia total potential CFGT (export and import) trade was around $30 billion USD in 2008. Truly, several nations were unable to meet their CFGT import demand in the period of global crisis started at the end of 2008; however, those countries were capable to raise CFGT import value of approximately $20 billion USD within Asia in 2008. The top potential CFGT importing country was South Korea, and its potential import value was around $15 billion USD in 2008, and the next was Pakistan ($3 billion USD).

Variation in the potential trade gap is observed among Asian nations. One of the major reasons is the variation of tariff rates of CFGT among Asian countries, regional trade agreements, etc. Other reasons may be lack of awareness and knowledge, insufficient technology, lack of skilled labour for production of CFGT, lack of trade facilities and infrastructure, etc.

8.3 Conclusion

This chapter examines the gravity equations considering the bilateral trade of CFGT export and import in the pre-global financial crisis period like 2005 and 2006 and focuses mainly on CFGT trade in 2008. The gravity model is used to explain determinants of export potential of CFGT for Asian nations within Asia and outside Asia such as in the North America and the European Union. This chapter estimates bilateral trade flows of CFGT and also its sub-categories like SPVS, CCT, WE and EEL applying the gravity model in Asia and observes its determinants. Income level, geographical distance and developmental position of both trading partners and country characteristics, economic policy reforms and available infrastructure are important determinants of CFGT trade and its sub-categories.

Potential trade gap is measured as the difference between predicted and actual trade among trading partners. Using the gravity model, this chapter measures the *potential export and import trade gap of climate friendly goods and technologies* in Asia in 2008. Through trade gap, this chapter estimates the value of trade opportunity of CFGT in Asia, identifies potential trading partners, and also suggests CFGT trade among the trade partners. The total estimated potential export of CFGT within Asia was nearly $32 billion US dollar (USD) in 2008. This study contributes in the empirical measurement of potential trade opportunity of CFGT for an individual country and also quantifies it for every trade partner. Trade opportunity of CFGT was more among Asian trading partners than outside Asia in 2008. It assists policymakers and governments in formulating appropriate trade and economic policy. It also helps negotiate trade in the right direction to tap the potential opportunity of CFGT export. It may stimulate CFGT export-led growth in Asia and also mitigates climate change issues.

There is a huge variation in the potential trade gap in CFGT among nations in Asia. One of the major reasons is the variation of trade restriction in Asian countries. Other reasons may be sociopolitical conditions and economic development policies which vary widely among Asian countries. The reasons for untapped potential export gap in CFGTs may be lack of awareness, unavailability of technology, lack of skilled labour for production of CFGT, unfavourable business environment, weak governance, inappropriate government policy towards CFGTs, lack of trade facilitations, etc. A more in-depth study of sub-regions is needed to explore these in detail. The next chapter focuses on South Asia region and highlights its possible potential trade opportunity.

Chapter 9
Emerging Climate Business in South Asia

Abstract This chapter analyses empirically the trade flow between countries and estimates correctly bilateral trade flows in South Asia. The total estimated potential export of CFGT in South Asia was nearly $15.5 billion USD in 2008. South Asian countries could increase their potential export of CFGT around $ 644.15 million USD in 2008. In terms of trade opportunity in 2008, India did better compare to other countries in South Asia. Yet, India could increase trade of $57.38 million USD within South Asia region. Pakistan ($375.56 million USD) was at the top followed by Sri Lanka ($211.2 million USD). India has captured CFGT market in South Asia and has emerged as an economic power in the region.

Keywords CFGT · Clean coal technology · Energy efficient technology · Wind energy · Solar photovoltaic system · Other code · Gravity model · Potential trade opportunity · Export · Import · GDP · GDP per capita · Distance · Trade gap · India · South Asia

9.1 Introduction

The Intergovernmental Panel on Climate Change (IPCC) report 2007 identified vulnerability in South Asia. In 2014 again the IPCC reaffirms vulnerability due to climate change and identifies South Asia as one of the most vulnerable regions in the world. In this context, can trade mitigate climate change? Does climate change stimulate to generate trade opportunity for climate friendly goods and technologies (CFGT) for nations in South Asia? Or, what is the trade opportunity in climate change for South Asian countries? Answering these questions is essential for searching trade policy strategy that mitigates climate change issues in South Asia.[1]

---

[1] In 2007 the IPCC reaffirms the climate change and identifies vulnerabilities in South Asia. Anthropogenic activities influence the physical and biological systems and consequences are inevitable. Key vulnerabilities include water resources, food supply, health, coastal settlements and some ecosystems (particularly the Sundarban mangrove forest in Bangladesh and India). South Asia is one of the most sensitive regions (see IPCC website, for details).

Advanced know-how and climate friendly technologies are readily available through liberalized trade,[2] which widely disseminate such technologies. Trade liberalization is good for the environment (Antweiler et al. 2001, Copeland and Taylor 2004, Liddle 2001). Free trade[3] has a contradictory impact on the environment, both raising pollution and motivating reduction of it. However, free trade should ensure availability of CFGT for developing countries where domestic industries are unable to produce CFGT in sufficient scale at affordable prices. So, free trade might provide access to CFGT for all. Capable countries might adapt such updated technologies and in turn these nations might be able to produce and export CFGT. Truly, trade might be a facilitator for driving technological innovations. Thus, trade could mitigate climate change through disseminating and exchanging updated technologies. Clean and low-carbon technologies improve energy efficiency and reduce climate change impact. Only a few developing countries, such as Brazil, China and Mexico, are important producers of clean energy technologies, while developing countries on the whole are net importers of environmental goods (World Bank 2008).

Most of the exporters of CFGT are from developed countries. A few exporters from developing countries became key players in the heat and energy management equipment, noise and vibration abatement, and in environmental services like air pollution control and solid waste management.[4] Few developing countries are among the top ten importers and exporters in various categories of environmental goods relevant to climate change mitigation (World Bank 2008). Though climate change is a threat, it also provides opportunity to redesign economic activities.[5] In this context developing countries must focus on production and trade of climate friendly goods and need more emphasis on it. Production structure of the economy tends to shift towards cleaner activities that generate less pollution. Brazil and Mexico are pioneers in the production of clean energy and/or green technologies (World Bank 2008).

This chapter analyses empirically the trade flow between countries applying the gravity model. The gravity model is able to estimate correctly bilateral trade flows making it one of the most stable empirical relationships in economics (Learner and

---

[2] For details, See World Bank 2008, Meyer-Ohlendorf and Gerstetter, 2009.

[3] Trade raises income level in developing economies and it will create demands for tighter environmental protection but lower trade barriers could hurt environment if heavy polluters move to countries with weaker regulations (Dinda 2004, Mukhopadhyay and Chakraborty 2005, Dean et al. 2009).

[4] See Veena Jha (2008, 2009) for more details.

[5] Over the past century, human activities have released large amounts of carbon dioxide (CO_2) and other greenhouse gases into the atmosphere. The majority of greenhouse gases come from burning fossil fuels to produce energy, although deforestation, industrial processes and some agricultural practices also emit gases into the atmosphere. In response to redesign human activities, a step change is needed, with action at global, national and local level. Local actors must be engaged in impact assessment and in identifying solutions. But global and national leadership is also required to manage the macro-scale effects that will accompany widespread efforts to adapt to climate change.

Levinson 1995). The distribution of CFGTs (goods or factors) across space is determined by gravity forces given the size of economic activities at each location. The gravity model is adopted to explain the role of economic size and endowments, distance between trading partner and membership of multilateral agreement, among others on trade of such climate friendly goods or/and sub-categories. In particular, this gravity study is a cross-sectional data analysis for estimating the gravity equation considering the bilateral total trade of the CFGT exports for the year 2008. The gravity model analysis is useful to explain determinants of export potential of CFGT for South Asian countries.

This chapter investigates the potential trade opportunity for CFGT in South Asia. This chapter also examines CFGT trade potentiality within region and interregions covering the European Union, North America and the rest of the world. Applying the gravity model, this chapter provides evidences with certain insights regarding potential untapped trade opportunity of CFGT in South Asia. This study might assist policymakers to form appropriate policy design on *climate change issues* and frame the mechanism to tap the emerging *trade opportunity* within South Asia[6] or with the rest of the world in the coming years.

9.2 Findings

Preliminary observations on CFGT export from South Asia are summarized in Tables 9.1 and 9.2. Table 9.1 displays South Asia's CFGT exports share of the world's CFGT exports during 2002–2009. South Asia's share of CFGT export in the world CFGT export is less than 1% of the Global CFGT export. Share of CFGT export of the world's total export increases slowly over time in South Asia region (see Fig. 9.1).

Table 9.2 and Fig. 9.2 show the trends of CFGT export and import shares of South Asia in the period of 2002–2009. The gap between CFGT import and export shares narrowed down during 2002–2007, and both nearly merged in 2008, and gap was only 0.07% in 2008 (CFGT import and export shares were 1.80% and 1.73%, respectively); however, the gap widens (3.25%) in 2009 due to the global crisis.

Table 9.1 South Asia's trade share of CFGT exports of the world's total CFGT exports during 2002–2009

Year	2002	2003	2004	2005	2006	2007	2008	2009
South Asia's share of CFGT export to the world's total CFGT exports	0.24	0.34	0.37	0.52	0.70	0.73	0.95	0.98

---

[6] Bangladesh, Bhutan, India, Maldives, Nepal, Pakistan and Sri Lanka belong to South Asia region. There is a regional block which is known as the South Asian Association for Regional Cooperation (SAARC). SAARC formed in 1985 with seven members – Bangladesh, Bhutan, India, Maldives, Nepal, Pakistan and Sri Lanka. Recently, Afghanistan has joined in SAARC.

Table 9.2 Trend of CFGT shares to South Asia's total export and import during 2002–2009

	2002	2003	2004	2005	2006	2007	2008	2009
Share of CFGT export to South Asia's total exports	0.32	0.63	0.72	0.86	1.17	1.31	1.73	0.23
Share of CFGT import to South Asia's total imports	1.48	1.35	1.40	1.56	1.55	1.83	1.80	3.48

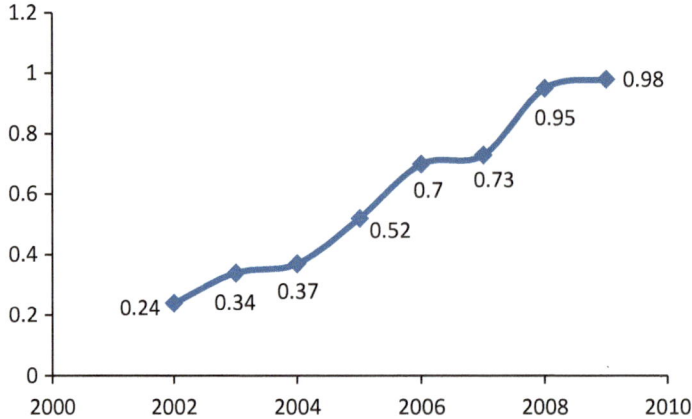

Fig. 9.1 Trend of South Asia's share of CFGT export to the world's total CFGT exports during 2002–2009

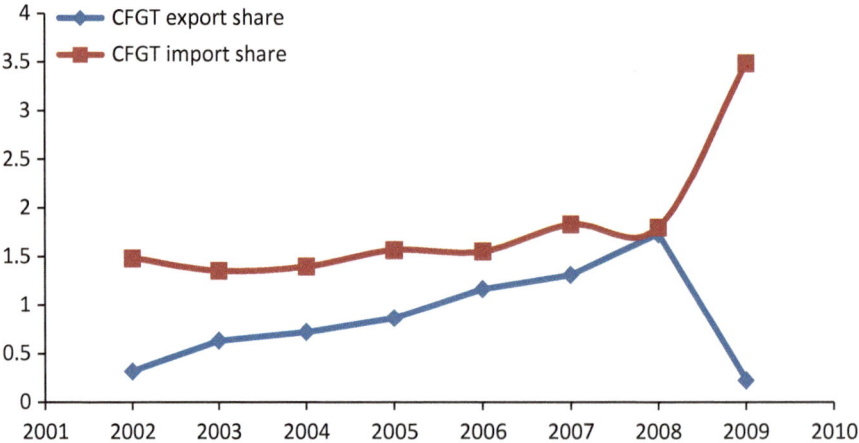

Fig. 9.2 Trend of CFGT trade shares in South Asia during 2002–2009

Table 9.3 provides the major region-wise destination of CFGT trade of South Asia (SAARC) in the period of 2002–2009. Major trade regions are South Asia, the European Union and the USA. Intraregional trade gap was 20% in 2002. Within South Asia CFGT import and export shares were 26.49% and 6.53% in 2002, respectively. Intraregional trade gap in South Asia was reduced to nearly 2.1% in 2008 but widened to 43% in 2009 (export and import shares were 3.8% and 1.51% in 2008 and 46.61% and 3.57% in 2009, respectively; see Table 9.3). Interregional trade gaps fluctuated and narrowed down to 5.3% in 2008 and widened to 26% in 2009 with EU, while with the USA, it was 10% in 2007 and 11.5% in 2009. Overall intraregional and interregional demands were nearly 1.5% and 98.5% for CFGT, respectively. Within the region CFGT demand was very poor, and it might be increased in the future to mitigate climate change issues. Intraregional demand for CFGT would be increased as economy develops with time and scope of CFGT trade might be increased.

Regional destinations of subcategories of CFGT trade in South Asia are also available in Table 9.3. Import of CCT and WE trade shares are significantly high with the EU. Import of CCT share from the USA is also significant. Import share of WE trade from the EU raised from 27% in 2002 to around 70% in 2007 and 2008. Export share of EEL dominates within South Asia and the EU. EEL import share was high as with 20% from the EU in 2003 and from South Asia in 2004, and after that it declined gradually and became the least in 2009. Import share of SPVS from the EU declined from 54% in 2002 to 30% in 2009 while its export share to the EU increased from 10% in 2002 to 67% in 2008 (see Table 9.4). It should be noted that SPVS export shares in the EU and the USA were zero, while it was high as 47% within South Asia in 2009. South Asia region has a good CFGT trade relation with the EU and the USA in the first decade of the twenty-first century.

Table 9.4 describes the summary statistics of major variables of South Asia in 2008, and the last column shows the pair correlation between import and other major explanatory variables with their significance levels. We examine the demand for CFGT which might be reflected in import of CFGT in South Asia. The demand for CFGT depends on income level and development position and access to market availability.

Table 9.5 Provides the estimated gravity model for CFGT import in South Asia in 2006. M1 of Table 9.5 is used as the base of the gravity model for CFGT import in South Asia in 2006. On the basis of statistical significance, the estimated base of gravity model for CFGT import in South Asia is

$$\ln M_{ij} = -36.9 + 0.884 \ln GDP_i + 1.367 \ln GDP_j + 0.389 \ln pcgdp_j - 1.66 \ln D_{ij} \quad (9.1)$$

M1 suggests that CFGT import directly depends on income levels of both trading countries and is inversely related to their distance. M2 adds regional agreement which is negative and highly significant. It suggests that regional agreement reduces CFGT import in South Asia in 2006. The estimated M2 is

Table 9.3 Region-wise destination of CFGT trade of South Asia during 2002–2009

CFGT

Year	South Asia		EU		USA	
	Import	Export	Import	Export	Import	Export
2002	26.49	6.53	20.16	27.99	6.65	24.67
2003	3.38	10.26	41.42	22.47	12.02	11.74
2004	3.13	8.07	40.81	23.72	10.90	14.20
2005	3.03	10.69	40.76	22.70	11.03	19.53
2006	2.31	6.97	39.56	20.95	9.35	31.55
2007	2.08	4.21	32.75	22.72	10.50	23.63
2008	1.51	3.80	34.77	29.44	6.84	18.26
2009	3.57	46.61	38.21	12.26	14.05	2.54

CCT

Year	Import	Export	Import	Export	Import	Export
2002	0.72		67.49		16.88	
2003	0.44	4.66	36.01	20.80	54.68	1.41
2004	1.20	0.76	48.78	8.50	18.06	20.66
2005	2.34	0.44	26.27	8.98	60.70	1.00
2006	8.44	4.95	32.59	19.83	30.46	7.60
2007	0.84	0.80	12.16	5.74	60.20	5.13
2008	0.66	28.17	31.51	2.44	16.15	1.84
2009	0.50		45.48		47.73	

EEL

Year	Import	Export	Import	Export	Import	Export
2002	15.46	72.71	19.27	4.61	0.05	12.51
2003	11.47	61.79	23.56	13.22	0.80	1.83
2004	21.94	53.84	19.38	16.66	1.24	0.47
2005	15.54	45.97	13.67	25.91	0.54	0.98
2006	10.21	36.91	8.96	22.48	0.63	0.27
2007	11.69	14.76	11.12	27.99	0.65	0.00
2008	4.49	8.84	8.83	24.57	0.57	0.00
2009	1.48	0.00	1.42	0.00	9.31	0.00

WE

Year	Import	Export	Import	Export	Import	Export
2002	14.71	30.48	27.34	0.00	3.62	0.00
2003	0.81	2.54	53.88	33.68	9.82	10.29
2004	0.57	1.55	56.43	48.47	8.97	13.29
2005	0.61	1.85	62.42	40.92	9.74	13.80
2006	0.62	2.00	64.67	36.28	8.26	18.66
2007	0.49	2.05	70.39	35.04	6.26	24.85
2008	0.16	0.64	70.16	35.10	7.48	21.82
2009	3.61	0.00	39.70	0.00	13.50	81.05

SPVS

Year	Import	Export	Import	Export	Import	Export
2002	10.54	23.17	53.95	10.71	1.82	10.49
2003	2.11	4.00	23.21	69.59	11.73	7.11
2004	2.31	5.14	30.68	67.34	10.83	10.18
2005	4.85	8.30	24.97	61.22	8.96	6.01
2006	3.19	7.07	24.64	49.25	6.49	9.97
2007	2.63	6.52	25.19	50.57	6.51	6.62
2008	0.86	3.57	34.23	67.08	3.60	2.64
2009	17.46	47.14	30.14	0.00	5.53	0.00

Note: Author's calculation (figures are percentage of South Asia's trade)

Table 9.4 Summary statistics of major variables in South Asia in 2008

Variable	N	Mean	S.D.	Min	Max	Import correlation with
Import ('000)	140	56108.6	250453.8	0.004	261,381	1
Tariffs weighted (%)	140	12.77	8.103	0.25	43.8	−0.32**
Distance (km)	140	5664.5	2606.1	374.7	14037.8	−0201
GDP reporter (USD)	140	$3.45 \times 10^{+11}$	$4.77 \times 10^{+11}$	$1.26 \times 10^{+9}$	$1.16 \times 10^{+12}$	0.493**
GDP partner (USD)	140	$1.31 \times 10^{+12}$	$2.54 \times 10^{+12}$	$1.37 \times 10^{+8}$	$1.41 \times 10^{+13}$	0.507**
PCGDP reporter (USD)	140	1644.8	1435.4	366.08	4134.93	−0.24**
PCGDP partner (USD)	140	28604.6	20341.2	366.08	109903.1	0.207*
Contiguous	127	0.055	0.229	0	1	0.335**
Common official language	127	0.1339	0.342	0	1	0.009
Colony	127	0.0236	0.152	0	1	−0.002
Regional agreement	127	0.0866	0.2824	0	1	0.106
lnPolicy reporter	68	1.4012	0.1683	1.253	1.589	0.38**
lnPolicy partner	126	1.5117	0.2233	0.9163	1.8083	0.14
lnInfrastructure reporter	68	1.2106	0.0149	1.194	1.224	−0.376**
lnInfrastructure partner	126	1.54836	0.311	0.642	1.887	0.247**
lnFDI reporter	127	6.4726	2.8293	2.48	9.738	0.617**
lnFDI partner	119	9.4277	1.3747	5.339	12.226	0.325**

Note: One and two stars indicate level of statistical significance at 5% and 1%, respectively

$$\ln M_{ij} = -27.2 + 0.928\ln GDP_i + 1.1645\ln GDP_j + 0.501\ln pcgdp_i$$
$$-2.228\ln D_{ij} - 2.724 D_{RegionalAgreement} \tag{9.2}$$

The estimated M3 is

$$\ln M_{ij} = -25.7 + 0.952\ln GDP_i + 1.133\ln GDP_j + 0.476\ln pcgdp_i$$
$$-0.683\ln pcgdp_j - 2.01\ln D_{ij} - 3.05 D_{RegionalAgreement} \tag{9.3}$$
$$+4.312 Infrastructure_j$$

The estimated M4 is

$$\ln M_{ij} = -29.7 + 0.945\ln GDP_i + 1.295\ln GDP_j + 0.555\ln pcgdp_i$$
$$-2.55\ln D_{ij} - 3.6 D_{RegionalAgreement} + 4.416 Policy_j \tag{9.4}$$

Partner's infrastructure and policy variables are added in M3 and M4 and both are positive and highly significant. However, in case of the presence of both in M5, only infrastructure is significant at low level. In M6 and M7, FDI is statistically

Table 9.5 Estimated gravity model for import of CFGT in South Asia in 2006

Variables	M1	M2	M3	M4	M5	M6	M7
Constant	−36.9*** (−6.58)	−27.2*** (−4.21)	−25.7*** (−4.16)	−29.7*** (−4.68)	−27.5*** (−4.3)	−24.9*** (−3.56)	−28.4*** (−4.05)
lnGDP_Reporter	0.884*** (9.53)	0.928*** (10.15)	0.952*** (11.08)	0.945*** (10.9)	0.952*** (11.09)	0.92*** (9.91)	0.938*** (10.69)
lnGDP_Partner	1.367*** (10.32)	1.1645*** (7.34)	1.133*** (7.58)	1.295*** (8.31)	1.2007*** (7.42)	0.953*** (3.76)	1.219*** (4.84)
lnpcgdp_reporter	0.38 (1.26)	0.501* (1.68)	0.476* (1.70)	0.555** (1.96)	0.5046* (1.80)	0.5002* (1.66)	0.512* (1.79)
lnpcgdp_partner	0.389* (1.65)	0.2883 (1.16)	−0.683** (−2.04)	−0.286 (−1.01)	−0.6408* (−1.90)	0.253 (0.99)	−0.53 (−1.48)
lnDistance	−1.66*** (−3.20)	−2.228*** (−3.61)	−2.01*** (−3.32)	−2.55*** (−4.23)	−2.188*** (−3.50)	−1.96*** (−3.19)	−2.15*** (−3.43)
DRegionalAgreement	–	−2.724*** (−2.33)	−3.05*** (−2.77)	−3.6*** (−3.18)	−3.331*** (−2.95)	−1.857 (−1.55)	−2.794** (−2.4)
lnInfrastructure_score partner	–	–	4.313*** (4.11)	–	3.013* (1.90)	–	2.097 (1.20)
lnPolicy_score partner	–	–	–	4.416*** (3.79)	1.91 (1.09)	–	2.628 (1.44)
Ln FDI partner	–	–	–	–	–	0.189 (0.68)	−0.038 (−0.14)
N	140	127	126	126	126	119	118
R²	0.6079	0.6111	0.6611	0.6542	0.6645	0.5981	0.6479
Adj R²	0.5933	0.5917	0.6410	0.6336	0.6416	0.5727	0.6185
Root MSE	2.4769	2.3567	2.2099	2.2324	2.2082	2.3247	2.1956

Note: Figures in parenthesis are t-values. '***', '**' and '*' denote the statistical level of significant at 1%, 5% and 10%, respectively

insignificant. FDI has no significant role in CFGT import in South Asia in 2006. Following fitting criteria M5 is the best fitted gravity model for CFGT import in South Asia in 2006.

The estimated M7 is

$$\ln M_{ij} = -28.4 + 0.938 \ln GDP_i + 1.219 \ln GDP_j$$
$$+0.512 \ln pcgdp_i - 2.15 \ln D_{ij} - 2.794 D_{RegionalAgreement} \tag{9.5}$$

Table 9.6 presents the estimated results of the gravity equation for CFGT export and import in 2008 in South Asia. Left and right sides of Table 9.6 provide results of CFGT export and import, respectively. From top to bottom, Table 9.6 has three parts having regression results, ANOVA and regression statistics. Bottom left part of Table 9.6 shows CFGT export model fitting criteria which suggest that overall fitting of South Asia's gravity equation is good in the cross-sectional data analysis in 2008. The R^2 value is 0.6093 that means variables used in the equation explained 60.93 per cent of the variations in CFGT export of South Asia. Adjusting R^2 0.5976 is also a good indicator after adjusting lost degree of freedom (df). Overall variation is judged through analysis of variance (ANOVA). Middle part of Table 9.6 displays

Table 9.6 Estimated gravity equation for export and import of CFGT in South Asia in 2008

Variables	Export 2008			Import 2008		
	Coefficients	Standard error	t Stat	Coefficients	Standard error	t Stat
Intercept	−124.03***	8.637	−14.36	−43.33***	10.993	−3.94
GDP_reporter	2.64***	0.141	18.73	0.91***	0.179	5.11
GDP_partner	0.92***	0.081	11.42	1.24***	0.107	11.56
pcgdp_reporter	6.84***	0.74	9.24	0.44	0.949	0.47
pcgdp_partner	−0.32***	0.11	−2.87	0.80***	0.151	5.32
Distw	−1.108***	0.254	−4.37	−2.02***	0.365	−5.54
Contig	0.38	1.05	0.36	2.68**	1.174	2.28
com_official language	0.17	0.69	0.25	1.95**	0.852	2.29
comlang_ethno	0.66	0.665	1.00	−1.36*	0.795	−1.71
Smctry	−1.13	1.44	−0.79	−2.58	1.592	−1.62
ANOVA						
	Sum of square	Mean sum of square	F stat	Sum of square	Mean sum of square	F stat
Regression	2634.34	292.7	52.15	2120.653	235.63	35.04
Residual	1689.36	5.6		1331.322	6.72	
Total	4323.7			3451.975		
Regression statistics						
R^2	0.6093			0.6143		
Adj.R^2	0.5976			0.5968		
RMSE	2.3691			2.593		
N	311			208		

Note: '***', '**' and '*' denote the statistical level of significant at 1%, 5% and 10%, respectively

ANOVA decomposing into explained regression and error residuals, corroborating their mean sum of squares and F statistic. Calculated F statistic 52.15 is highly significant with (9, 301) degree of freedom.

Table 9.6 provides the point estimation of coefficients of the gravity equations. Statistically significant coefficients are highlighted with stars marked. The coefficients of GDP reporter, GDP partner, per capita GDP of reporter and that of partner and geographical distance between pair countries are statistically significant at 1% level. Other variables are statistically insignificant. Considering only statistically significant coefficients, the estimated export of CFGT in 2008 in South Asia is

$$\ln X_{ij} = -124.03 + 2.64 \ln GDP_i + 0.92 \ln GDP_j$$
$$+6.84 \ln pcgdp_i - 0.32 \ln pcgdp_j - 1.108 \ln DT_{ij} \tag{9.6}$$

The export elasticity of CFGT with respect to GDP of reporting country is highly elastic while it is inelastic with partner country. This suggests that CFGT export would be increased by more than 2.64% if income of the reporting country increases by 1%. CFGT export elasticity with respect to partner's GDP is 0.92 which indicates that CFGT export increases by 0.92% in reporter country as partner country's GDP increases by 1%. In other words, it can be interpreted that demand for CFGT increases with rising partner's income level and thereby South Asia's CFGT export increases definitely with less than 1%. Truly one part of partner country's internal demand may be fulfilled by CFGT production in reporting countries (here, South Asian nations). CFGT export with respect to reporting country's economic development (per capita GDP) is highly elastic (6.84). CFGT export increases by 6.84% as per capita GDP rises 1% in South Asia in 2008. The export elasticity of CFGT with respect to partner countries' per capita GDP (development) is inelastic (−0.32). CFGT export decreases by 0.32% as GDP per capita rises 1% in partner country. It is clear from these findings that CFGT export increases with income (GDP) while declines with per capita income in the rest of the world.

CFGT export in South Asia directly depends on economic growth of both and reporter's development, whereas it is inversely related with partner's development.[7] Rising global awareness about climate change and related international policy strategies might increase demand for CFGT as well as its production opportunity in South Asia. Truly, CFGT trade might expand with rising opportunity of green business in South Asia. Distance is inversely related with trade as per gravity model. CFGT export elasticity with respect to distance is −1.108, which is elastic. Trade is inversely related with geographical distance between pair of countries. Reducing transportation cost definitely increases bilateral trade. Constant term is statistically significant and indicates to add other factors or variables.

Bottom right part of Table 9.6 shows CFGT import model fitting criteria of gravity equation in South Asia in 2008. The R^2 is 0.6143 and indicates that used vari-

---

[7] South Asian Preferential Trade Agreement (SAPTA) is working in general trade, not much for CFG trade. South Asia should review if it captures potential trade opportunities.

ables in the equation explained 61.43% of the variations in CFGT import in South Asia. Adjusting R^2 0.5968 is a good indicator after adjustment of df. ANOVA helps to judge overall variations of model and decomposes into explained and error parts. Right middle part of Table 9.6 shows ANOVA. Calculated F 35.04 is significant with df (9, 198).

Top right part of Table 9.6 provides the point estimation of coefficients of CFGT import gravity equations in South Asia in 2008. Almost all coefficients are statistically significant except reporter's per capita GDP and small country. The coefficients of GDP reporter and partner, per capita GDP of partner, are positive and statistically highly significant, while geographical distance between pair countries is negative and statistically significant. Contiguity and common official language is significant at 5% level, and common ethno is significant at 10% level. Perhaps it may make easy communication between trading countries. Considering only statistically significant coefficients, the estimated CFGT import in South Asia in 2008 is

$$
\begin{aligned}
\ln M_{ij} = {} & -43.33 + 0.91 \ln GDP_i + 1.24 \ln GDP_j + 0.44 \ln pcgdp_i \\
& + 0.80 \ln pcgdp_j - 2.02 \ln DT_{ij} + 2.68 D_{Contig} \\
& + 1.95 D_{Com Official\_Lang} - 1.36 D_{Com\_lang\_Ethno}
\end{aligned}
\tag{9.7}
$$

CFGT import elasticity with respect to GDP of reporting country is 0.91 which is inelastic, while it is elastic (1.24) with partner country. This suggests that CFGT import might be increased by 0.91% as income of the reporting country increases by 1%. CFGT import elasticity with respect to partner's GDP is 1.24 which suggests that CFGT import increases by 1.24% in reporter country as partner country's GDP increases by 1%. It can be interpreted that demand for CFGT increases in reporting country with rising partner's income level. CFGT import elasticity with respect to partner country's economic development is inelastic (0.80). CFGT import increases by 0.80% as partner's per capita GDP rises 1% in South Asia. It is clear from these findings that CFGT import increases with income and economic development in the rest of the world.

This chapter is based on Dinda (2011a, b, and 2014a, b) and provides certain insights regarding trade opportunity of CFGT in South Asia. South Asian nations are far below the expected trade performance as per definition of *potential trade gap*[8] in the literature (Baldwin 1994, Nilsson 2000, Egger 2002 and Dinda 2014a, b). Applying the gravity model, we estimate the predicted export value of the reporting country with its trade partners.[9] The total estimated potential gap in the export of CFGT in South Asia in 2008 is around $ 644.15 million USD. 'Trade potential gap in CFGT export' suggests the possible scope to increase the CFGT export with its partner. Hence, trade opportunity value of CFGT export is $ 644.15 million USD in South Asia in 2008. This trade gap suggests that South Asian nations could increase the export of CFGT nearly $ 650 million USD.

---

[8] In this study the 'trade potential gap in CFGT' is the difference of predicted and actual trade.

[9] It should be mentioned that here statistically significant coefficients are non-zero and insignificant coefficients are zero. Hence, only statistically significant coefficients are used to predict the export value.

9.3   Conclusion

Using the gravity model, this chapter estimates the value of trade opportunity of CFGT in South Asia and identifies potential bilateral trading partners. The total estimated potential export of CFGT in South Asia was nearly $15.5 billion USD in 2008. This study contributes in the empirical measurement of potential trade opportunity of CFGT for individual country with their trading partners and suggests for possible expansion of CFGT trade in 2008. South Asian countries could be increased their potential export of CFGT around $ 644.15 million USD in 2008. In terms of trade opportunity in 2008, Pakistan ($375.56 million USD) was at the top followed by Sri Lanka ($211.2 million USD), India ($57.38 million USD), etc. Pakistan, Sri Lanka and India may prefer to do trade in CFGT regionally.

The above analyses provide the actual position of each country with respect to trade of CFGT and its sub-categories. Socio-political conditions and economic development policies varied widely among South Asian countries. Wide variation in trade gap might be due to cross-border trade restrictions in the region. The reasons for untapped potential export gap in CFGT might be the lack of education and awareness, unavailability of technology, lack of skilled labour for production of CFGT, unfavourable business environment, weak governance, inappropriate government policy towards CFGTs, lack of trade facilitations and infrastructure, etc. These findings would assist the policymakers and/or the government to formulate economic policy regarding CFGT and negotiate trade partners in the right direction to tap the potential CFGT trade opportunities in South Asia.

From South Asia region, India emerges as an economic power. Next chapter focuses on India's trade opportunity in 2008 and also highlights its CFGT trade applying trend analysis in the period of 2003–2017.

Chapter 10
Climate Change and CFGT Trade in India: A Case Study

Abstract This chapter shows the trends of CFGT export and import during 2003–2017. India's CFGT export to South Asia grows rapidly around at 3.5% growth rate in postcrisis period (2010–2017) compared to 2.1% in precrisis period (2003–2007). Using gravity model, this study quantifies trade opportunities of CFGT in India in the precrisis and during global crisis periods. Potential trade gap in CFGT trade in India was around $6 billion USD in 2008. With these it also identifies constraints and helps to expand capacity and/or strengthen its capability in the advancement of capturing new opportunities in production and trade in CFGT.

Keywords CFGT · Clean coal technology · Energy-efficient technology · Wind energy · Solar photovoltaic system · Other code · Gravity model · Potential trade opportunity · Export · Import · GDP · GDP per capita · distance · potential trade gap · India · Look East policy

10.1 Introduction

In this twenty-first century, climate change emerges as a new challenge for the economic development in emerging economy like India. During industrialization developed nations used huge fossil fuel energy. Accumulated fossil fuel consumption has contributed a lot to change the global climate. Contribution of less developed countries is negligible (compared to industrially developed nations) or little to cause climate change; however, they are facing its consequences (Dinda 2014a, b, 2015). LDCs have low capacity to adapt to the harsh impacts of climate change. Now, LDCs have already several hindrance or obstacles for their developmental activities; now climate change emerges as an additional constraint. However, climate change provides also certain opportunity to grow with new products such as climate friendly goods and technology (CFGT) or environmental goods and services (EGS), which are considered to be material, equipment or technology which may address particular environmental problem or climate change issues (Nguyen

© The Author(s), under exclusive licence to Springer Nature Switzerland AG 2019
S. Dinda, *Climate Friendly Goods and Technologies in Asia*, SpringerBriefs in
Environmental Science, https://doi.org/10.1007/978-3-030-02475-8_10

and Kalirajan 2015). One product is said to be environmentally good, which is less harmful to environment, which is environmentally preferable to similar or near products. EGS or CFGT help to measure, prevent or minimize environmental damage or correct climate changes (OECD 1999).

Now, questions may arise as follows: Can trade promote to mitigate climate change in emerging economies? Does climate change create any business opportunity in emerging economy like India? What is the trade opportunity for CFGT in India? Who are the potential trade partners of emerging India within South Asia region, in Asia and the world? This chapter attempts to realize India's CFGT export and import and quantify trade opportunities of CFGT in India. With these it also identifies constraints and helps to widen capacity and strengthen its capability in the advancement of capturing new opportunities in production and trade in CFGT.

Climate change creates limitation for development in one hand and also provides, on other hand, opportunity to grow with newly clean and CFGT. Truly, it creates the pace for opportunity to rearrange or redesign the economic activities. Trade is the engine of growth in the supply-driven economy. Trade certainly promotes developing countries through generating export earnings and accessing the updated technologies and adapting them in their economic system. Trade might mitigate climate change through disseminating and exchanging the low-carbon technologies. The main objective of the clean technology is to improve energy efficiency and minimize environmental impacts. The products associated with certain clean technologies that have relatively less adverse impact on the environment are the main focus of this study. This chapter examines the potential trade in CFGT in India during 2003–2017. This chapter includes also postcrisis period and updated data till 2017 from UNCOMTRADE (downloaded six-digit HS data on 28th of July 2018). This study provides evidence on CFGT trade opportunity in India in the period of 2003–2017 and suggests formulating the policy on *climate change and trade* for mitigating climate change issues in regional and global levels.

This study highlights the export potential trade of CFGT in India. CFGT trade was non-existent in India in 1990s; however, in post-global financial crisis, India emerges as a promising market for CFGT. CFGT might be a newly large industry focusing business on equipment or system supply like renewable energy plant, portable water treatment, noise and vibration abatement, etc. (Monkelbaan 2011). This chapter deals with the potential trade of India's CFGT within Asia, with the European Union (EU), North America (the USA and Canada) and the rest of the world. This study is mainly based on the application of the partial gravity model[1] in 2008. This analysis is useful to explain determinants of India's export potential of CFGT within Asia, the EU and the USA. Gravity model is adopted to explain the role of economic size and endowments, distance between trading partners and membership of multilateral agreement, among others on trade of such climate friendly goods and technology and its sub-categories. In particular, the gravity anal-

---

[1] True gravity model is not suitable and applicable for single economy. Part of the gravity equation is estimated after dropping reporter country's GDP and its income per capita, which are common for all bilateral trading partners.

ysis considers the bilateral CFGT trade of imports and exports in India in 2008. This chapter provides both cross-sectional data analysis for estimating the gravity equation in the crisis year 2008 and time series simple trend analysis for the period of 2003–2017.

10.2 Findings in the Crisis Year 2008

This part of analysis is a follow-up of earlier chapters. Overall trade performance was quite satisfactory in Asia and especially in India in crisis year 2008. Asia's actual export of CFGT trade[2] was nearly $119.74 billion USD in 2008. Correspondingly India's actual trade value of CFGT export was nearly $3.55 billion USD in 2008. It was 1.95% of India's total trade to world in 2008.

In the earlier section, we have discussed about region-wise India's CFGT export destinations. Now we estimate the above said gravity model for selected year 2005 and 2008 and analyse the results (see Table 10.1). Here, GDP and per capita GDP of the reporting country, India, are dropped because of no variation in a year, such as 2005 or 2008. In 2005, coefficients of partner's GDP and per capita GDP and distances between trading partners are statistically significant at 1% level. India's CFGT export raises with partner's GDP while it declines with partner's development (per capita GDP). Small country and common colony are significant at 5% and 10% level, respectively. CFGT export is more common in colony countries compared to others, while in case of small countries, it falls due to low demand for CFGT in 2005.

The estimated coefficients are almost similarly significant except constant term in 2008 (Table 10.1). Constant term is highly significant in 2008, not in 2005. Magnitude of the estimated coefficient of partner's GDP is more than one. It suggests that it is elastic and more sensitive in 2008 compared to 2005. Similarly the coefficient of distance is less sensitive in 2008 compared to 2005. In terms of model fitting criteria, gravity model in 2008 is better fitted than that in 2005.

As per statistically significant coefficients, the estimated India's export of CFGT equation in 2005 is

$$\ln X_{ij} = 0.984 \ln GDP_j - 0.38 \ln pcgdp_j - 1.068 \ln DT_{ij} + 0.84 D_{common\,colony} - 3.79 D_{smctry} \tag{10.1}$$

Considering statistically significant coefficients, the estimated equation of CFGT export of India in 2008 is

---

[2] Out of it, intraregional and interregional trades were 61.19 and 58.55 billion USD, respectively. Intraregional demand was nearly 51% and only 49% for interregional demand of CFGT. It is true that internal demand within Asia is very high for the climate friendly goods, and over time it will increase with economic development

Table 10.1 Estimated gravity model for export of CFGT of India in 2005 and 2008

	Export in 2005	Export in 2008
Constant	−5.03	−10.275***
	(−1.53)	(−3.66)
lnGDP partner	0.984***	1.196***
	(10.93)	(15.49)
lnPer capita GDP partner	−0.3845***	−0.638***
	(−3.31)	(−5.8)
lnDistance	−1.0678***	−0.7015***
	(−3.71)	(−2.72)
Contiguity	1.43	0.597
	(1.26)	(0.59)
Common office language	−0.042	−0.305
	(−0.06)	(−0.5)
Common ethno	0.402	0.746
	(0.61)	(1.32)
Colony	1.19	−0.259
	(0.62)	(−0.15)
Common colony	0.8434*	0.748*
	(1.86)	(1.83)
Small country	−3.79**	−3.12**
	(−2.15)	(−1.98)
R^2	0.5528	0.6784
Adj.R^2	0.5265	0.6593
RMSE	1.8415	1.6459
N	163	162

Note: Figures in parentheses are t-value. '***', '**' and '*' denote the statistical level of significant at 1%, 5% and 10%, respectively.

$$\ln X_{ij} = -10.275 + 1.196\ln GDP_j - 0.638\ln pcgdp_j$$
$$-0.7015\ln DT_{ij} + 0.748D_{common\,colony} - 3.12D_{smctry} \qquad (10.2)$$

It is interesting that the significant variables in 2005 are also significant in 2008 and difference is a constant term which is significant in 2008. Constant term is statistically significant which might capture other unknown factors in India. Detailed depth study is required to explore the reasons behind it.

Using these estimated gravity models, we can estimate the potential export of CFGT in India in 2005 and 2008. 'Potential trade gap' is measured as the difference between actual export and predicted value of export of CFGT in this study as mentioned in the earlier chapter. 'Potential trade gap' is a measurement of performance of a bilateral trade flow relative to the model predicted mean trade value for India for a given year. Using the gravity model, we estimate the predicted export trade value of India with its trade partners. For the analysis purpose, this study mainly focuses on the quantification of 'potential trade gap' in India. 'Potential trade gap of CFGT' itself suggests that there is a scope to increase the export of CFGT with respective trading partners.

Now this chapter highlights the potential trade of CFGT in India. Using estimated export gravity equation in Asia, we also show the estimated potential export gap in Figs. 10.1 and 10.2 for 2005. Potential export gap is measured and displays graphically for all trade partners of India. Figures 10.1 and 10.2 show the trade gaps for countries in Asia-Pacific region and the European Union, respectively. In Figs. 10.1 and 10.2, the horizontal line is the benchmark line and bars indicate trade gaps. These bars are standardized trade gaps. Bars below the benchmark line show that actual trade of CFGT is less than estimated potential trade. In other words, bars below the benchmark indicate the untapped trade opportunity for India in CFGT trade.

Total estimated export potential trade gap of CFGT in India is nearly 6 billion USD in 2008. This trade gap suggests that India could increase the export of CFGT around $6 billion USD in 2008.

Following gravity equation, total estimated potential CFGT export was $9.536 billion USD in India in 2008, while actual export was only $3.55 billion USD. Actually India utilized only 37.2% of its potential export trade of CFGT in 2008. India could increase export of CFGT by 62.8% in 2008. India can utilize

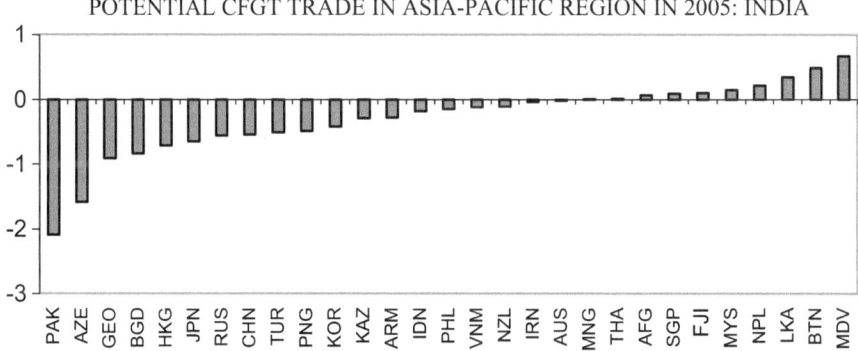

Fig. 10.1 India's trade opportunity in Asia and Pacific in 2005

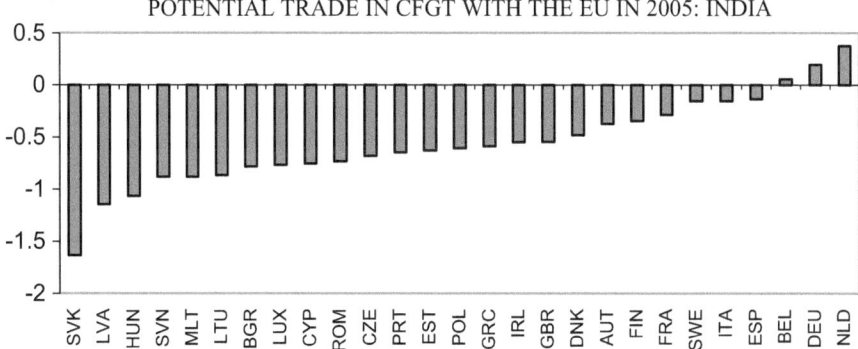

Fig. 10.2 India's trade opportunity in the EU in 2005

moderately trade of CFGT and has potential to increase its trade opportunity in CFGT. Roughly total potential export gap of CFGT in India was $6 billion USD in the world out of $4.9 billion USD in Asia in 2008. Definitely it suggests increasing trade with respective partners.

From Fig. 10.3 it is clear that India's potential trade is huge in Asia and Asia-Pacific region. Within Asia, India could increase the CFGT export to Pakistan, Mongolia, Bangladesh, Armenia, Kazakhstan, Azerbaijan, Japan, Vanuatu, Russia, China, Kyrgyz Republic, Hong Kong, Korean Republic, Indonesia, Iran and the Philippines.

Figure 10.4 displays that India has a great potential export trade of CFGT to developed countries. The most important and encouraging India's CFGT trade partners are Luxembourg, the UK, Latvia, Cyprus, Greece, Hungary, Slovenia, Slovakia, Austria, Finland, Ireland, Poland, Spain, Lithuania, Bulgaria, Romania, Denmark, Sweden, France, Italy and the Czech Republic. India has trade potential to increase trade of CFGT with Canada.

The estimated India's CFGT export potential gap in 2008 was around $5 billion USD within Asia and $1.01 billion USD with the EU. India's export potential trade gap of CFGT was higher in Asia than the EU. India has strong trade potential with

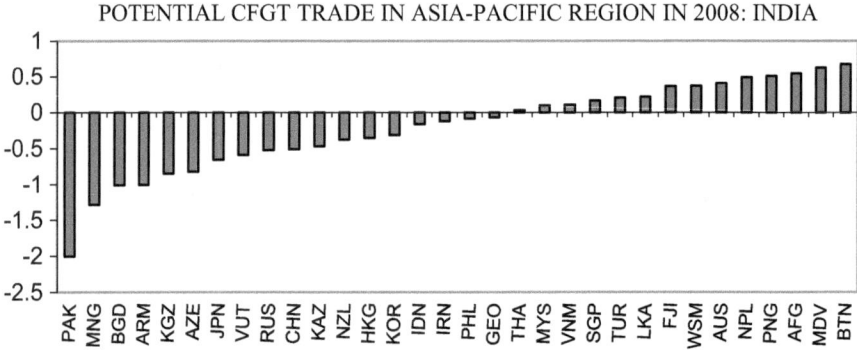

Fig. 10.3 India's trade opportunity in Asia and Pacific Region in 2008

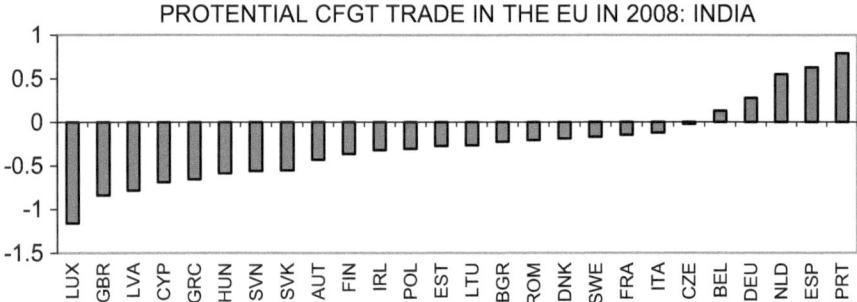

Fig. 10.4 India's trade opportunity in the EU in 2008

Pakistan, Bangladesh, China, Japan, Russia and South Korea, and estimated potential export gap of CFGT to these countries was nearly $4.9 billion USD. India's CFGT export potential gap to Pakistan and Bangladesh was $4.4 billion USD. India should explore this potential trade after revising its *Look East Policy* and could stimulate to control climate change in the region.

India's CFGT potential trade top partners in the EU were the UK, France, Italy, Poland, Greece and Austria, and the potential trade gap was nearly $1 billion USD in the crisis period. India had potential to increase its export of CFGT to Asia and the EU approximately more than $6 billion USD in 2008. There was a huge variation in the potential trade gap among nations due to lack of knowledge, backdated technology, lack of effective skilled labour and entrepreneurs, lack of trade facilitations and infrastructure, etc. Do these factors change over time and corresponding effect on trade in India? Truly cross-sectional study could not capture this dynamics. To capture time dynamics, it is an essential need to examine a trend analysis to overview trade performance in India. Next we study trend analysis for the period of 2000–2017.

10.3   Trend Analysis in the Period of 2000–2017

Initially we observe positive trends of overall export and import in India that are measured in terms of percentage of GDP in the early twenty-first century for period of 2000–2015. India's export reached at peaks at 24.27% and 25.43% in 2008 and 2013, respectively. Figure 10.5 displays the trends of trade in India during 2000–2015, while Fig. 10.6 shows the trends of CFGT export and import (measured in billion US dollar) in India during 2003–2017. It is clearly visible that CFGT export increases rapidly during precrisis period (2003–2008) and rises slowly in the post-crisis period (2009–2017). CFGT import follows a similar pattern to CFGT export till 2008; however, import rises rapidly after 2014. CFGT import is more than that of export in India during 2003–2017, and CFGT trade gap increases after 2014.

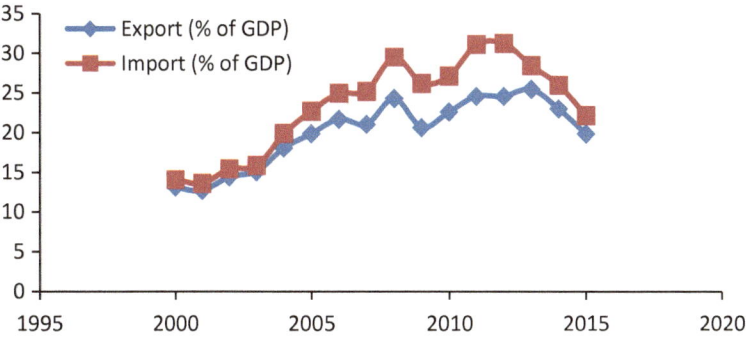

Fig. 10.5 Trends of export and import trade (% of GDP) in India during 2000–2015

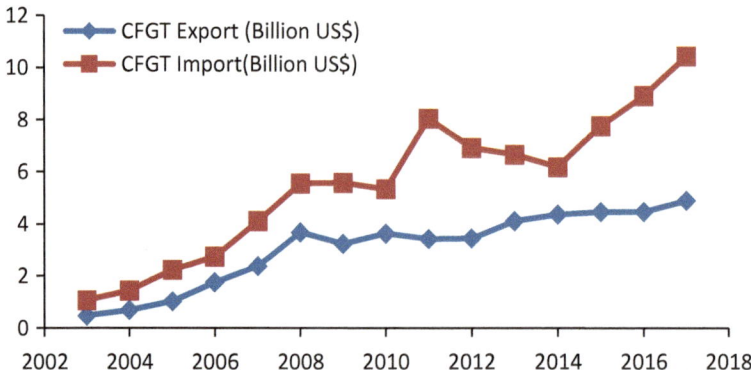

Fig. 10.6 Trends of CFGT export and import in India during 2003–2017

Table 10.2 Share of CFGT export and import in India during 2003–2017

Year	Share of CFGT export	Share of CFGT import
2003	0.809	1.484
2004	0.917	1.457
2005	1.036	1.604
2006	1.463	1.561
2007	1.651	1.904
2008	2.056	1.824
2009	1.914	2.136
2010	1.681	1.565
2011	1.195	1.786
2012	1.204	1.449
2013	1.241	1.463
2014	1.380	1.376
2015	1.700	2.043
2016	1.721	2.577
2017	1.666	2.373

Table 10.2 displays the trend of shares of CFGT export and import in India in the period of 2003–2017. In 2003, CFGT export share was less than 1% of India's total export, while CFGT import share was 1.48% of India's import. CFGT export share reached at peak at 2.05% in 2008 and declined to 1.67% in 2017. However, CFGT import share touched at high at 2.13% in 2009 and declined to 1.37% in 2014 and gained momentum after 2014 and reached at peak at 2.58% in 2016 (see Fig. 10.7).

From Fig. 10.7, it is clear that India's CFGT export share shows a cycle during 2003–2017. It has three major phases: (i) CFGT export share increased at faster rate during 2003–2008, (ii) it declined in the period of 2009–2011 and (iii) recovery started in 2012 and improved slowly during 2012–2017. Both CFGT export and import share started to recover after 2014; however, CFGT import share recovered at a faster rate than its export.

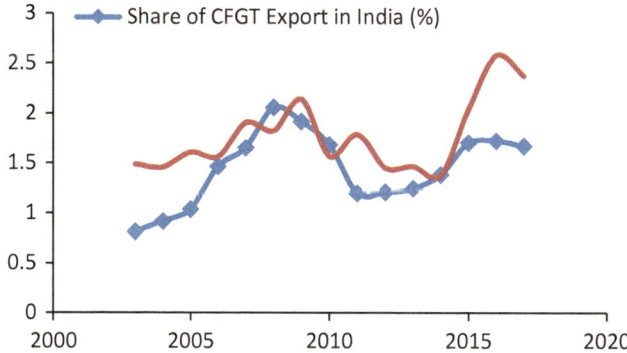

Fig. 10.7 Trends of shares of CFGT export and import in India during 2003–2017

Table 10.3 Major region-wise CFGT export and import (billion US $) in India in 2003, 2005, 2008, 2012 and 2016

Destinations	Trade	2003	2005	2008	2012	2016
EU	Export	0.113724	0.255206	1.09234	0.617542	0.985031
	Import	0.535595	1.089974	2.231958	2.230582	1.753054
US	Export	0.052889	0.21159	0.650346	0.581526	0.702283
	Import	0.143167	0.27131	0.386497	0.680322	0.539844
Asia	Export	0.205621	0.326368	1.017836	1.19316	1.792884
	Import	0.322521	0.779661	2.649123	3.703722	6.313784
South Asia	Export	0.03476	0.054084	0.126209	0.185255	0.277685
	Import	0.010939	0.018206	0.052981	0.04441	0.049915

Note: Author's calculation

Now we focus on certain regional destination for CFGT trade in India during 2003–2017. Table 10.3 provides major regional destinations of India's CFGT export and import measured in billion US dollar for selected years such as 2003, 2005, 2008, 2012 and 2016.

India has imported CFGT mainly from Asia and the European Union. In 2003 and 2005, India imported more CFGT from the EU and exported to Asia; however, it reverses after the global financial crisis. In 2016, India's CFGT import rapidly increased from Asia and correspondingly that declined from the EU. Actually, India has shifted its import destination from the EU in precrisis period (2003–2007) to Asia in postcrisis era (2010–2017). India's CFGT import from the USA rises slowly in postcrisis period except 2016 and its export to the USA is more or less stable in postcrisis era. Figure 10.8 presents bar diagram for selected regions for selected years. Compared to major regional destinations of CFGT trade, South Asia region was not significantly visible in 2003 and 2005 in Fig. 10.8. India has rapidly increased its CFGT export in South Asia from $ 0.03476 billion USD in 2003 to $ 0.278 billion USD in 2016 and 0.333 billion USD in 2017 (see Table 10.3 and

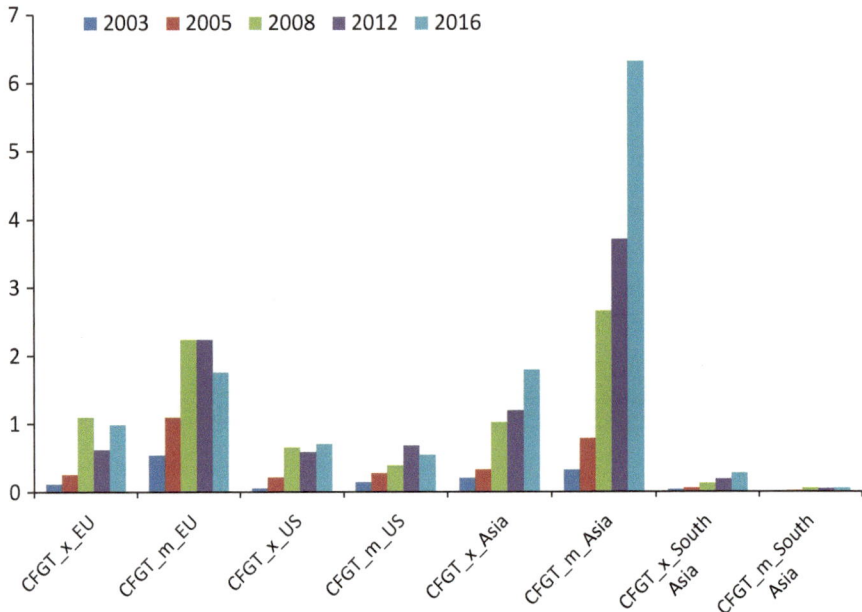

Fig. 10.8 Regional destinations of CFGT export and import in India in 2003, 2005, 2008, 2012 and 2016

Table 10.4 India's CFGT export (billion US dollar) to South Asia during 2003–2017

Year		2003	2004	2005	2006	2007	2008	2009
CFGT X (billion US $)		0.035	0.047	0.054	0.067	0.075	0.126	0.085
Year	2010	2011	2012	2013	2014	2015	2016	2017
CFGT X (billion US $)	0.139	0.194	0.185	0.220	0.254	0.259	0.278	0.333

Note: Author's calculation

Table 10.4). India has certain positive role in South Asia to adapt and mitigate climate change issues.

Now we investigate in detail India's CFGT export trade growth in South Asia region during 2003–2017. The whole period is also divided into two periods – pre- and postcrisis period, i.e. 2003–2007 and 2010–2017, respectively. Pre- and postcrisis period are marked as red and green colours in Fig. 10.9. India's CFGT export to South Asia increased at a slower rate in the precrisis period of 2003–2007 (see dotted line, $y = 0.010 \times -20.18$, in Fig. 10.9) compared to higher rate in postcrisis period of 2010–2017 (see dashed line, $y = 0.024 \times -48.55$, in Fig. 10.9). The slope of the dashed line is steeper than that of dotted line, which indicates a switchover from slower rate in precrisis period to higher rate in postcrisis period. In this context, exponential curve might be suitable for Indian's CFGT export to South Asia in the entire period of 2003–2017.

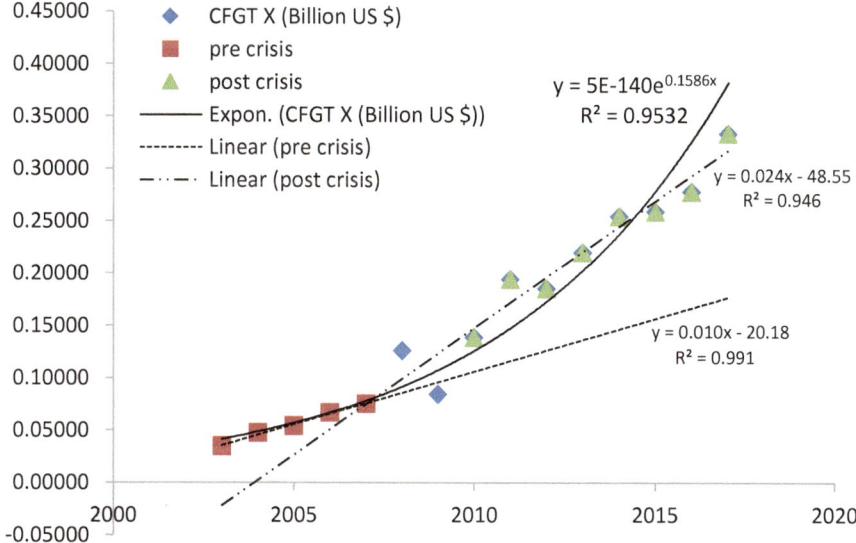

Fig. 10.9 India's CFGT export to South Asia during pre- and postcrisis in the period of 2003–2017

Fig. 10.10 CFGT export growth in India during 2003–2017

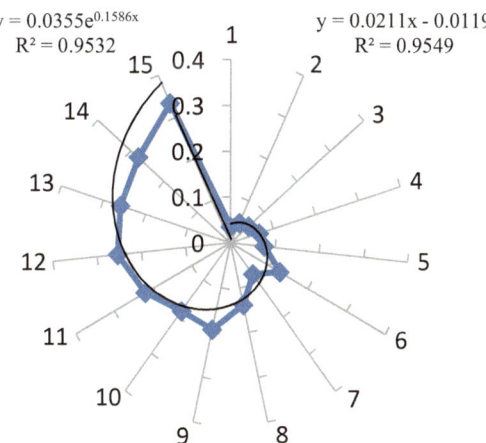

Table 10.4 displays India's CFGT export in billion US dollar to South Asia during 2003–2017. Observing the period of 2003–2009 in Table 10.4, it is clear that CFGT export amount of India to South Asia is below $ 0.10 billion except in 2008 ($ 0.126 billion), which is higher than $ 0.10 billion. Amounts of CFGT export to South Asia are above $0.10 during 2010–2012 and above $0.20 billion in the period of 2013–2017. Now, it is a need to investigate CFGT export in India in 2008. It is discussed in the next section in detail.

Year-to-year change of India's CFGT export to South Asia is segmented linear growth; however, actually, CFGT export of India to South Asia is growing exponentially in the entire period of 2003–2017 (see Fig. 10.10). India is stepping strong footing in South Asia to promote CFGT to mitigate climate change issues in local, regional and global level.

10.4   Conclusion

The paper highlights the estimated trade gap of CFGT in India in 2008 and trend of CFGT trade in the period of 2003–2017. Applying the gravity model, this paper measures the *potential trade gap* and suggests possible expansion of the export trade of CFGT among trading partners. The total estimated export potential trade gap of CFGT in India was around $6 billion USD in 2008. This study contributes in the empirical measurement of potential trade of CFGT in India and quantifies potential trade gap of individual partners. It supports the possible emergence of CFGT export-led growth in India and also mitigates climate change problems in the future. India might adopt few policies to improve and raise CFGT production and trade. The reasons for untapped potential export gap in CFGT might be the lack of awareness, unavailability of technology, lack of human capital for CFGT production, the government policy, lack of trade facilitations, etc. Our next agenda is to collect more information and explore these in details using updated econometric tools and forecast potential CFGT trade for 2020, 2030 and 2050. More depth study is needed to overcome these limitations.

CFGT import of India increases gradually over time in the period of 2003–2017. India's CFGT import share rises in each interregional level. It was high in the EU in precrisis period while it increases at a faster rate in Asia especially after global financial crisis. India's CFGT export share has increased in South Asia, Asia and the EU in the postcrisis period. India's CFGT export to South Asia grows around 3.5% in postcrisis period (2010–2017) compared to 2.1% in precrisis period (2003–2007).

Chapter 11
Mapping the Movement of CFGT

Abstract This chapter provides mapping of one-to-one possible trade opportunity in Asia. It identifies bilateral trade flow between pair of countries and indicates possible future trade movements in selected countries in Asia such as China, South Korea, Japan, Indonesia, the Philippines, Thailand, India, Pakistan and Sri Lanka. Here this study considers trade gaps in 2005 and 2008 and highlights possible direction in bilateral trade flow between pair of trading partners.

Keywords CFGT · SAARC · ASEAN · APTA · Gravity model · Potential trade opportunity · Export · Import · Potential trade gap · Asia · China · Japan · India · South Asia · Sri Lanka · Pakistan · South Korea

Now this chapter discusses the potential trade opportunity of CFGT and focuses on possible future trade movement of selected countries in Asia such as China, South Korea, Japan, Indonesia, Malaysia, the Philippines, Thailand, India, Pakistan and Sri Lanka. Here this study considers trade gaps in 2005 and 2008 and highlights possible direction in bilateral trade between pair of trading partners. In the last part of this chapter, we display bar diagrams showing trade opportunities of selected countries in 2005 and 2008 in Asia and the EU. Figures in graphs indicate bilateral trade gaps and compare it between 2005 and 2008. Now, we mainly focus on trade partners within Asia and the EU and discuss possible trade opportunities for the said regions.

Japan has a comparative advantage in technology and is able to increase its potential trade in CFGT. The estimated potential CFGT exports of Japan were $175 million US dollar within Asia and $24 million USD with the EU in 2008. Japan had the strongest trade potential with Russia, and the estimated potential export of CFGT was nearly $133 million USD in 2008. Japan has the potential to increase its export of CFGT to Asia and the EU. The estimated total potential export of CFGT was approximately $200 million USD in 2008. Within Asia, Japan has strong export potential in CFGT to Afghanistan, Armenia, Azerbaijan, Bangladesh, Bhutan, Georgia, Kazakhstan, Kyrgyz Republic, Mongolia, Nepal, Russia and Sri Lanka.

Japan's most important and encouraging CFGT potential trade partners in the EU are Finland, Greece, Latvia, Lithuania and Romania.

China properly utilized CFGT trade in 2008; however, still China had potential CFGT trade gap in 2008. Within Asia, China has strong trade potential to export to Armenia, Bhutan, Hong Kong, Nepal and South Korea. The European Union is the most important destination of China. China should explore this potential trade and might stimulate to control climate change.

Within Asia, *South Korea* has strong trade potential in CFGT export to Afghanistan, Azerbaijan, Japan, Kyrgyz Republic and Nepal. One observation should be noted that South Korea did not much encourage its CFGT potential trade with the EU, except Luxembourg in 2008. South Korea's estimated potential CFGT export was $1.043 billion US dollar within Asia in 2008. South Korea had the strongest trade opportunity with Japan, and the estimated potential CFGT export with Japan was nearly $1.042 billion USD in 2008. South Korea could explore it to tap this potential trade gap.

Indonesia's estimated potential export in CFGT in 2008 was $43.8 million USD and $27.26 million US dollar in Asia and the EU, respectively. Indonesia's export opportunity for CFGT was higher in Asia than the EU. China was the strongest potential trade partner of Indonesia, and the estimated potential CFGT export to China was nearly $27.2 million USD in 2008. Indonesia's CFGT potential top trade partners in the EU were Austria, Denmark, Italy, Poland, Spain, Sweden and the UK, and the potential trade was nearly $16.2 million USD. Indonesia's total potential trade was approximately more than $71 million USD in 2008. Within Asia, Indonesia has the trade potential of CFGT export to Kazakhstan, Azerbaijan, Russia, Turkey, China, Iran, Pakistan, South Korea, Afghanistan and Bangladesh. Indonesia's most important and encouraging CFGT potential trade is with the EU, especially Austria, Bulgaria, Cyprus, the Czech Republic, Denmark, Estonia, Finland, Ireland, Italy, Latvia, Poland, Romania, Slovenia, Spain, Sweden and the UK.

Thailand has potential trade opportunity of CFGT with Asian nations such as Afghanistan, Azerbaijan, Iran, Kazakhstan, Kyrgyz Republic, Mongolia, Russia, Turkey, etc. Thailand can increase its CFGT trade with the EU especially with Eastern European countries like Estonia, Latvia, Lithuania, Slovenia, Slovakia, etc.

The Philippines can raise CFGT trade within Asia potential trading partners which are Bangladesh, China, Hong Kong, India, Nepal, Pakistan, Russia, South Korea, Turkey, etc. The Philippines have certain trade advantage due to its geographical location in Asia-Pacific area. Within the Asia-Pacific region, potential trading partners are Australia, Bangladesh, China, Fiji, Hong Kong, India, Nepal, New Zealand, Pakistan, Russia, South Korea, Turkey, etc. Its trade opportunity of CFGT with Asian nations is limited compared to Asia-Pacific region. This is because of its location. The Philippines might raise its CFGT trade with the EU especially with countries like Austria, Denmark, Finland, Italy, Romania, Slovakia, Spain, Sweden, the UK, etc.

Now this chapter will discuss the potential export trade opportunity of CFGT for India, Pakistan and Sri Lanka in South Asia.

Sri Lanka can increase its potential export trade of climate friendly goods and technology. Within Asia, Sri Lanka has strong potential export of CFGT to Kazakhstan, Indonesia, Iran, Malaysia, Mongolia, Pakistan, the Philippines, Singapore and Thailand. Sri Lanka could also increase the CFGT trade with Canada. The most important and encouraging Sri Lanka's CFGT potential trade is with the European Union, especially Austria, Cyprus, Denmark, Hungary, Latvia, Romania and Spain. The estimated Sri Lanka's potential exports of CFGT were 425,000 US dollar within Asia and 177,000 US dollar with EU in 2008. Sri Lanka has potential to increase export of CFGT within Asia and EU.

Pakistan has trade potential to increase CFGT trade of 17.5 million US dollar with the USA and Canada in 2008. The estimated Pakistan's CFGT exports potential was 893.39 million US dollar within Asia and 65.79 million USD with the EU in 2008. Pakistan's export potential trade in CFGT is more within Asia than any other region. Pakistan has the strongest trade partner in terms of export potential with India, and estimated export of CFGT to India was nearly 838.7 million USD in 2008. Pakistan should explore this potential trade and might stimulate to control climate change in the South Asia region. Pakistan's CFGT potential trade top partners with the EU were France, Germany, Italy and the UK; and potential trade was nearly 55.49 million USD in 2008. Pakistan has potential to increase its export of CFGT to Asia and the EU, the USA and Canada. Pakistan has a great potential to increase its trade potential particularly in CFGT. Within Asia, Pakistan has strong trade potential in CFGT export to Azerbaijan, Bangladesh, China, Hong Kong, India, Indonesia, Iran, Japan, Kazakhstan, South Korea, Kyrgyz Republic, Malaysia, Nepal, Russia, Singapore, Thailand, Turkey and Vietnam. Pakistan has a great trade potential in CFGT trade with developing countries. The most important and encouraging Pakistan's CFGT potential trade is with the European Union, especially Austria, Belgium, Cyprus, the Czech Republic, Denmark, France, Finland, Germany, Greece, Hungary, Ireland, Italy, Lithuania, the Netherlands, Poland, Portugal, Romania, Slovakia, Spain, Sweden and the UK.

India has the potential to increase its trade opportunity in CFGT trade. Within Asia, India has potential trade partner in CFGT to Central Asia, South Asia, East Asia and Southeast Asia. India's most important and encouraging CFGT trade partners are in the European Union and North America. Most potential emerging bilateral trading partners are China, Pakistan, Japan, Bangladesh, Indonesia, Australia, Austria, Bulgaria, Canada, Denmark, Finland, France, Italy, Luxembourg, Poland, Romania, Russia, South Korea, Spain, Sweden, the USA and the UK. India's potential trade in CFGT is higher in Asia than the EU in postcrisis period. India had strong trade opportunity of CFGT with Pakistan, Bangladesh, China, Japan, Russia, and South Korea, and the estimated potential export of CFGT to these countries was nearly $4.9 billion USD in 2008. India's CFGT export potential to Pakistan and Bangladesh alone was $4.4 billion USD in 2008. India could explore this potential trade with revising the *Look East Policy* and stimulate to control climate change in the region. India's CFGT potential top trade partners in the EU were the UK, France, Italy and Austria, and the potential trade was nearly $1 billion USD in 2008.

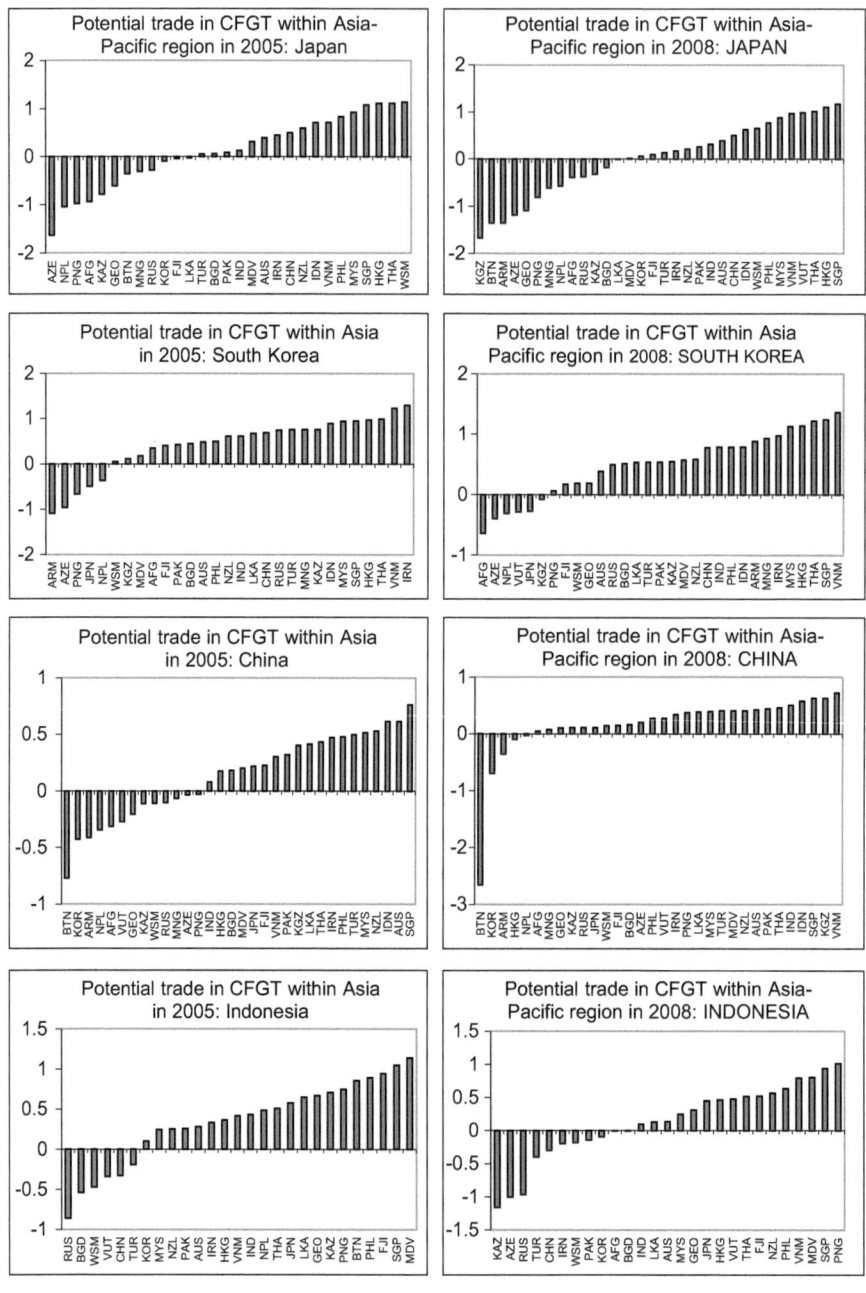

Fig. 11.1 Comparative trade gaps of selected countries in Asia within Asia-Pacific region in 2005 and 2008

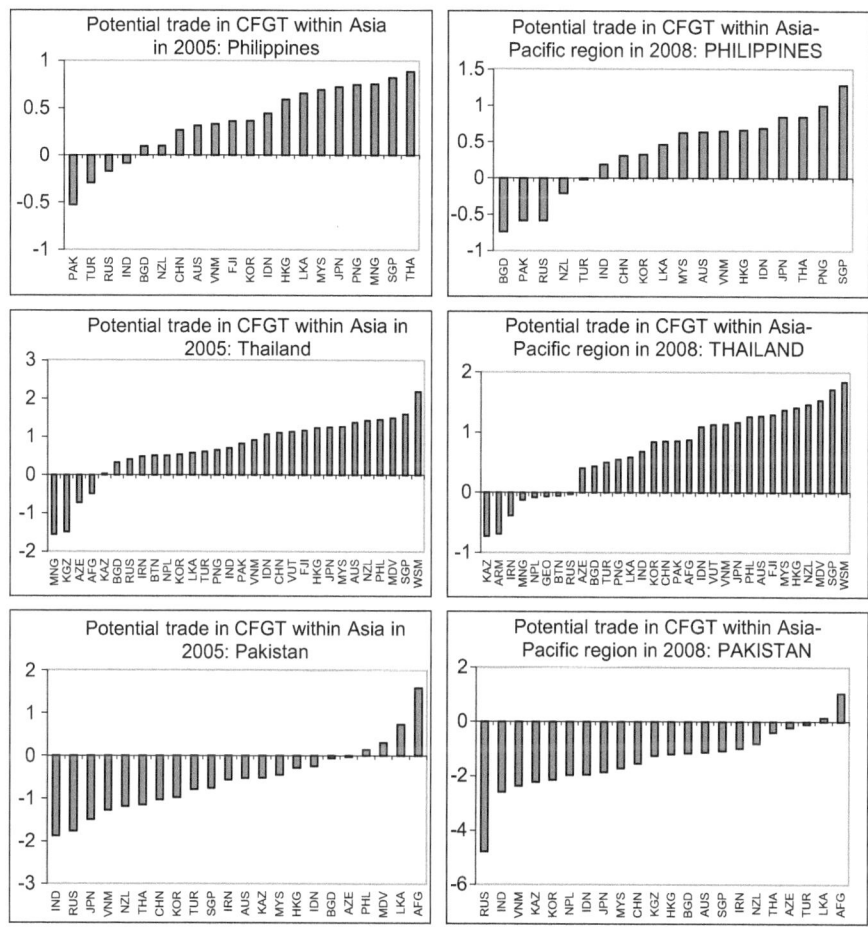

Fig. 11.1 (continued)

Now, potential trade gaps between trading partners are calculated for selected countries in Asia and map possible trade opportunities for each of them. This study compares the estimated potential trade gaps of selected countries between 2 years for 2005 and 2008. The estimated trade gaps are shown in bar diagram in Figs. 11.1 and 11.2. Bar lines (in Figs. 11.1 and 11.2) display the potential trade gap between reporter and partner countries separately for the years of 2005 and 2008. Here, this chapter shows the mapping and movement of trade opportunities within Asia and the EU between 2005 and 2008. Figure 11.1 provides comparative trade gaps of selected countries within Asia in 2005 and 2008. Similarly, Fig. 11.2 shows comparative trade gaps of selected countries in Asia with partner countries in the European Union in 2005 and 2008.

So, individual and regional partners of selected countries in Asia are identified in this study. This study provides mapping of one-to-one possible trade opportunity in Asia. It will help to form adaptive policy and settle the bilateral agreement with potential trading partners.

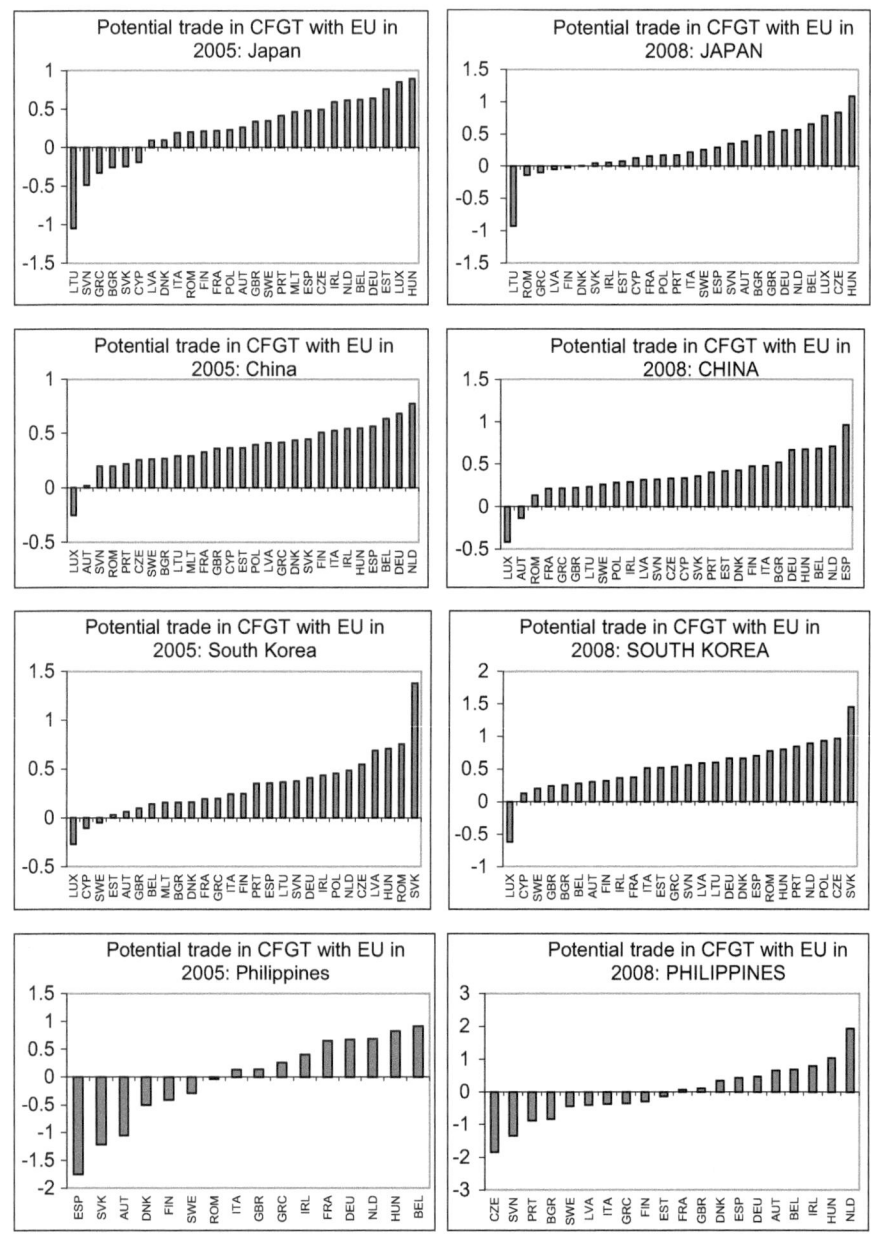

Fig. 11.2 Comparative trade gaps of selected countries in Asia with countries in the European Union in 2005 and 2008

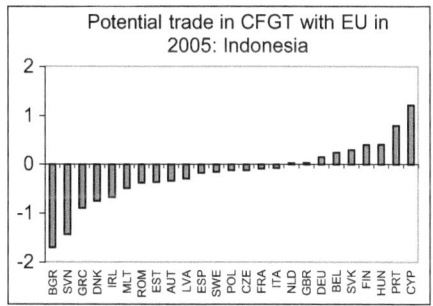

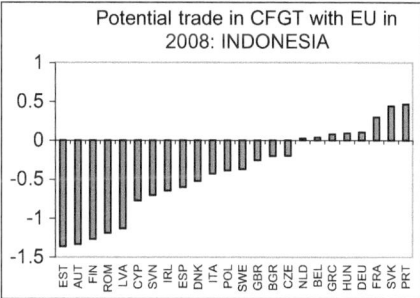

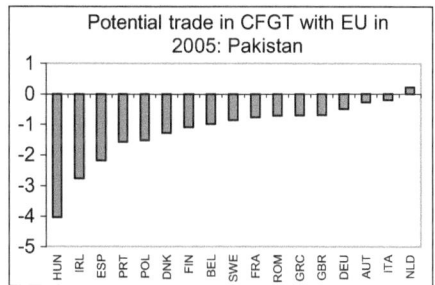

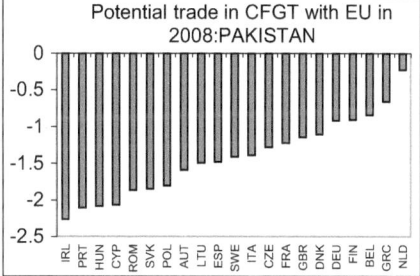

Fig. 11.2 (continued)

Chapter 12
Conclusion

Abstract This chapter concludes with remarks. Most of countries in Asia were importer of CFGT in early part of this century, and later some of them turned to become exporter of it. Japan was exporter and importer of CFGT from the beginning of the twenty-first century. Japan was the lead exporter of EEL and later joined China, Hong Kong, Taiwan, Macao, etc. Countries in Asia mostly imported SPVS and wind energy technology from the EU and Japan. Later, China exported SPVS to Asia and the world. Gravity analysis provides determining factors of CFGT trade in Asia. Economic size, distance, level of development and country characteristics are important determining factors for CFGT trade in Asia. Factors beyond the border such as economic reforms, policy, transparency, regulations and infrastructure are most significant factors in determining trade patterns in CFGT among nations in Asia. However, tariffs of CFGT are negligible and less significant among other factors beyond the border like reform policy, regulation and infrastructure in both trading partners. Trade liberalization definitely promotes clean technology transfer and provides channel to mitigate climate change issues in global scale.

Keywords CFGT · Clean coal technology · CCT · Energy-efficient lighting · EEL · Wind energy · WE · Solar photovoltaic system · SPVS · Other codes · Gravity model · Trade opportunity · Export · Import · GDP · Potential trade gap · Trade liberalization · Technology transfer · Climate mitigation · Beyond the border · Policy reform · Regulation · Infrastructure

This study examines trade pattern and performance of climate friendly goods and technology in Asia during 2002–2017 highlighting the period of Asian crisis in 1997–1998 to the global financial crisis in 2008–2009. Asian crisis in 1997–1998 did not affect the USA and the world economy, while the US-led global financial crisis in 2008–2009 (which was originated from the USA) severely affected the entire world. Truly, the crisis of developed economy transmitted to the rest of the world through trade channel. It also affected trade in CFGT. In the global financial crisis 2008, the share of CFGT in the world export was 2.55% in 2008. The share of CFGT export in the world varied from 2.3% in 2002 to 2.8% in 2009 to 6.1% in

S. Dinda, *Climate Friendly Goods and Technologies in Asia*, SpringerBriefs in Environmental Science, https://doi.org/10.1007/978-3-030-02475-8_12

2017. The import share varied from 2.2% in 2002 to 2.7% in 2009 to 6.5% in 2017. CFGT trade share improved from Kyoto Protocol 1997 to Copenhagen Summit 2009. This study briefly discusses on trade diversity in the direction of CFGT in the post Kyoto Protocol in 1997. This study, initially, assesses trade performance of CFGT, analyses and estimates it in terms of trade gravity model and finally calculates potential trade opportunities for selected countries in Asia, mainly focusing on the period of 2002–2008.

This study examines trade performance of CFGT using trade indices like export and import shares, competitiveness index, RCA, Michelaye index and regional orientation index for Asia and regional trade blocks for the periods of 2002–2008. Most of the countries in Asia were importer of CFGT in early part of this century, and later some of them turned to become exporter of CFGT. Asia imported CFGT mostly from the USA and the EU in starting years of the twenty-first century. From 2004 onwards, some countries in Asia started to export CFGT to other nations within Asia. Japan was both exporter and importer of CFGT from beginning of this century. Japan was the lead exporter of energy-efficient lighting, and later China, Hong Kong, South Korea, Taiwan, Macao, etc., from East Asia and Indonesia, Malaysia, Vietnam and Thailand from Southeast Asia started to export EEL to Asia-Pacific region, the EU, North America and rest of the world. Countries in Asia mostly imported SPVS and wind energy technology from the EU and also from Japan. In later phase, China exported such technologies to the rest of Asia and the world. South Asia initially imported such technologies from the USA and the EU, but after 2004–2005, specifically, its import direction of EEL and SPVS shifted to East and Southeast Asia. After the world financial crisis in 2008, the directions of export and import of CFGT or its sub-categories changed drastically, and emerging countries in Asia (like China and India) diversified their trade at a faster rate. CFGT import share in Asia was more than that of export in the early twenty-first century, and export share improved later.

It is clear that Japan and China were at top exporter list of all categories of CFGT except CCT in the said study periods. Countries in top three exporters for at least one sub-category in Asia were Malaysia, Singapore and Thailand. Japan, China and South Korea were at top two positions of CFGT import, while Hong Kong, Singapore, Russia, India and Turkey were top next importers of at least one sub-category. In 2009, China became top exporter in SPVS and EEL and top importer in SPVS and wind power; Japan was top exporter in wind power and CCT and top importer in EEL. Malaysia was top three in exporting SPVS following Japan. India became the third CCT exporter following Singapore in 2009.

Regional bias for CFGT is observed in Asia following ROI values exceeding one. Countries in ASEAN and APTA had no comparative advantage; however, they were net importers from countries within Asia. Some countries in SAARC like Pakistan, Sri Lanka and India would like to trade in CFGT regionally. No region or no country in Asia had a comparative advantage in clean coal technologies in 2002. SAARC had shown interest in production and trade of CCT in 2006 onwards. Pakistan and Singapore were only countries in 2008 who have secured more than one in RCA. India was at third rank with RCA value of 0.85. It suggests that SAARC countries had developed expertise in CCT in 2008.

Michelaye index is also worked out for ASEAN, APTA and SAARC for CFGT and its sub-categories. Positive values come only for energy-efficient lighting. Thailand, Vietnam and Macao in APTA performed better in terms of their export pattern. Solar photovoltaic systems had positive figures for ASEAN indicating the comparative advantage in SPVS during 2002–2008. All other categories had negative figures indicating that ASEAN as a group did better in terms of export pattern to its own import structure for SPVS. ASEAN as a group had regional bias towards its own region for all codes except SPVS in 2002 and 2008. SAARC as a group had regional bias for EEL in both years and CCT in 2008. APTA as a group has regional bias for OC and CCT in 2008. Asia is bias for OC in 2002 and 2008.

The above analysis based on trade indices indicates the factual position of each country with respect to trade of CFGT and its sub-categories. Gravity analysis provides us the determining factors of CFGT trade in Asia. Economic size, distance, level of development, country characteristics, endowments and tariffs are important determining factors for trade in CFGT in Asia. However, factors beyond the border such as economic reforms, policy, transparency, regulations and infrastructure are listed as significant factors in determining trade patterns in CFGT between countries in Asia. However, tariffs of CFGT are less significant among other factors beyond the boarder like reform policy, regulation and infrastructure in both trading partners.

Applying the econometric technique, this study estimated the trade gravity model and found the predicted trade value of CFGT. Using these predicted estimation and actual trade of CFGT, this study has calculated trade gap for each pair of trading partners during 2002–2009; however, this study has focused more in 2005 and 2008. The estimated trade gap in CFGT was around $35 billion US dollar in Asia only; in other words, the value of loss of potential trade in CFGT in Asia was $35 billion US dollar in 2008. Truly, trade gap in CFGT in 2005 in Asia was nearly $16 billion US dollar which was much less than that of in 2008. However, potential trade opportunity in CFGT improves over time, particularly after the Copenhagen Summit 2009. Emerging economies like China and India also agree to reduce carbon emission within 2030. The Copenhagen Summit 2009 has reached a global consensus and agreed for low-carbon economy. This study also identified bilateral potential trade movement of CFGT within and outside Asia.

On the basis of trade performance of Asia in trade of cleaner technologies in the first decade, this study provides policy suggestion for global consensus for low-carbon economy for moving towards greener future. Global and regional collective efforts are required to step up to address climate change focusing more on renewable and clean energy and energy efficiency. The transition to a low-carbon economy requires a stable long-term carbon pricing policy with appropriate regulatory regime. The effective carbon pricing policies should be focused on competitiveness and job creation and promote climate friendly technologies and innovation having significant reduction of emissions. Renewable energy sector should emerge as an engine of economic growth. Recently, India is in the right path with setting targets for solar and wind energy production in large scale. Apart from renewable energy, eco-friendly construction is essential for preparing cities with need of incorporation of green technology into their designs of infrastructure services including sanitation. There are also growing opportunities in climate friendly financial solutions with a wide range of green bonds to microloans for entrepreneurs. In this context, India is ahead of the world and set target to zero pollution. India emphasizes on global cooperation focusing on climate change issues. Considering climate as global public good, every nation should adopt mitigation policy accordingly at regional and national levels.

Appendix

Table A.1 List of 64 climate friendly goods and technology

Serial no.	HS codeS6 digit(2002)	Definition
1	380210	Activated carbon
2	392690	Articles of plastics & arts. of oth. mats. of 39.01–39.14, n.e.s. in Ch.39
3	392010	PVC or polyethylene plastic membrane systems to provide an impermeable base for landfill sites and protect soil under gas stations, oil refineries, etc. from infiltration by pollutants and for reinforcement of soil
4	560314	Nonwovens, whether or not impregnated, coated, covered or laminated, of man-made filaments; weighing more than 150 g/m2 for filtering wastewater
5	701931	Thin sheets (voiles), webs, mats, mattresses, boards and similar nonwoven products
6	730820	Towers and lattice masts for wind turbine
7	730900	Containers of any material, of any form, for liquid or solid waste, including for municipal or dangerous waste
8	732111	Solar-driven stoves, ranges, grates, cookers (including those with subsidiary boilers for central heating), barbecues, braziers, gas-rings, plate warmers and similar non-electric domestic appliances and parts, thereof, of iron or steel
9	732190	Stoves, ranges, grates, cookers (including those with subsidiary boilers for central heating), barbecues, braziers, gas-rings, plate warmers and similar non-electric domestic appliances and parts, thereof, of iron or steel
10	732490	Water-saving shower
11	761100	Aluminium reservoirs, tanks, vats and similar containers for any material (specifically tanks or vats for anaerobic digesters for biomass gasification)

(continued)

S. Dinda, *Climate Friendly Goods and Technologies in Asia*, SpringerBriefs in Environmental Science, https://doi.org/10.1007/978-3-030-02475-8

Table A.1 (continued)

Serial no.	HS codeS6 digit(2002)	Definition
12	761290	Containers of any material, of any form, for liquid or solid waste, including for municipal or dangerous waste
13	840219	Vapour generating boilers, not elsewhere specified or included hybrid
14	840290	Superheated water boilers and parts of steam generating boilers
15	840410	Auxiliary plant for steam, water and central boiler
16	840490	Parts for auxiliary plant for boilers, condensers for steam, vapour power unit
17	840510	Producer gas or water gas generators, with or without purifiers
18	840681	Turbines, steam and other vapour, over 40 MW, not elsewhere specified or included
19	841011	Hydraulic turbines and water wheels of a power not exceeding 1,000 kW
20	841090	Hydraulic turbines and water wheels; parts, including regulators
21	841181	Gas turbines of a power not exceeding 5,000 kW
22	841182	Gas turbines of a power exceeding 5000 kW
23	841581	Compression type refrigerating, freezing equipment incorporating a valve for reversal of cooling/heating cycles (reverse heat pumps)
24	841861	Compression type refrigerating, freezing equipment incorporating a valve for reversal of cooling/heating cycles (reverse heat pumps)
25	841869	Compression type refrigerating, freezing equipment incorporating a valve for reversal of cooling/heating cycles (reverse heat pumps)
26	841919	Solar boiler (water heater)
27	841940	Distilling or rectifying plant
28	841950	Solar collector and solar system controller, heat exchanger
29	841989	Machinery, plant or laboratory equipment whether or not electrically heated (excluding furnaces, ovens, etc.) for treatment of materials by a process involving a change of temperature
30	841990	Medical, surgical or laboratory stabilizers
31	848340	Gears and gearing and other speed changers (specifically for wind turbines)
32	848360	Clutches and universal joints (specifically for wind turbines)
33	850161	AC generators not exceeding 75 kVA (specifically for all electricity generating renewable energy plants)
34	850162	AC generators exceeding 75 kVA but not 375 kVA (specifically for all electricity generating renewable energy plants)
35	850163	AC generators not exceeding 375 kVA but not 750 kVA (specifically for all electricity generating renewable energy plants)
36	850164	AC generators exceeding 750 kVA (specifically for all electricity generating renewable energy plants)
37	850231	Electric generating sets and rotary converters; wind-powered
38	850680	Fuel cells use hydrogen or hydrogen-containing fuels such as methane to produce an electric current, through an electrochemical process rather than combustion

(continued)

Table A.1 (continued)

Serial no.	HS codeS6 digit(2002)	Definition
39	850720	Other lead acid accumulators
40	853710	Photovoltaic system controller
41	853931	Discharge lamps, (e.g. ultraviolet), fluorescent
42	854140	Photosensitive semiconductor devices, including photovoltaic cells whether or not assembled in modules or made up into panels; light-emitting diodes
43	900190	Mirrors of other than glass (specifically for solar concentrator systems)
44	900290	Mirrors of glass (specifically for solar concentrator systems)
45	903210	Thermostats
46	903220	Manostats
47	700800	Multiple-walled insulating units of glass
48	730431	Tubes, pipes & hollow profiles (excl. of 7304.10–7304.29), seamless, of circular cross-section, of cold-drawn/cold-rolled (cold-reduced) steel
49	730441	Tubes, pipes & hollow profiles (excl. of 7304.10–7304.39), seamless, of circular cross-section, of stainless steel, cold-drawn/cold-rolled (cold-reduced)
50	730451	Tubes, pipes & hollow profiles (excl. of 7304.10–7304.49), seamless, of circular cross-section, of alloy steel other than stainless steel, cold-drawn/cold-rolled (cold-reduced)
51	840682	Steam turbines & oth. vapour turbines (excl. for marine propulsion), of an output not >40 MW
52	841012	Hydraulic turbines & water wheels, of a power > 1000 kW but not >10,000 kW
53	841013	Hydraulic turbines & water wheels, of a power > 10,000 kW
54	850239	Electric generating sets n.e.s. in 85.02
55	850300	Parts suit. for use solely/princ. with the machines of 85.01/85.02
56	850440	Static converters
57	902830	Electricity meters, incl. calibrating meters therefor
58	903020	Cathode-ray oscilloscopes & cathode-ray oscillographs
59	903031	Multimetres
60	903039	Instruments & app. for meas./checking voltage/current/resistance/power (excl. of 9030.31), without a recording device
61	890790	Floating structures other than inflatable rafts (e.g. rafts (excl. inflatable), tanks, coffer-dams, landing-stages, buoys & beacons)
62	847989	Machines & mech. appls. having individual functions, n.e.s./incl. in Ch.84
63	842129	Filtering/purifying mach. & app. for liquids (excl. of 8421.21–8421.23)
64	842139	Filtering/purifying mach. & app. for gases, other than intake air filters for int. comb. engines

References

Anderson, J.E. 1979. A Theoretical Foundation for the Gravity Equation. *American Economic Review* 69 (1): 106–116.

Anderson, J.E., and E. van Wincoop. 2003. Gravity with Gravitas: A Solution to the Border Puzzle. *American Economic Review* 93 (1): 170–192.

———. 2004. Trade Costs. *Journal of Economic Literature* 42 (3): 691–751.

Anderson, M., M. Ferrantino, and K. Schaeffer. 2005. Monte Carlo Appraisals of Gavity Model Specifications, *US International Trade Commission Working Paper* 2004–05-A.

Antweiler, Werner, Brian R. Copeland, and M. Scott Taylor. 2001. Is Free Trade Good for the Environment? *American Economic Review* 91 (4): 877–908.

Baier, Scott L., and Jeffrey H. Bergstrand. 2001. The Growth of World Trade: Tariffs, Transport Costs, and Income Similarity. *Journal of International Economics* 53 (1): 1–27.

Balassa, B. 1965. Trade liberalization and revealed comparative advantage. *The Manchester School of Economic and Social Studies* 33 (1): 99–123.

———. 1966. Tariff reductions and trade in manufactures among the industrial countries. *American Economic Review* 56 (3): 466–473.

———. 1977. Revealed' comparative advantage revisited: An analysis of relative export shares of the industrial countries, 1953–1971. *The Manchester School of Economics & Social Studies* 45 (4): 327–344.

———. 1979. Incentive policies in Brazil. *World Development* 7 (11,12): 1023–1042.

———. 1986. Dependency and trade orientation. *The World Economy* 9 (3): 259–274.

Balassa, B., and L. Bauwens. 1987. Intra-industry specialisation in a multi-country and multi-industry framework. *Economic Journal* 97 (388): 923–939.

Baldwin, R.E. 1994. *Towards an Integrated Europe*. London: Centre for Economic Policy Research.

Baldwin, R.E., and Taglioni, D. 2006. Gravity for Dummies and Dummies for Gravity Equations, *NBER Working Paper* No. W12516.

Balineau, G., and J. de Melo. 2011. *Stalemate at the negotiations on environmental goods and services at Doha round*. Ferdi Working Paper no 28.

Balistreri, E.J., and R.H. Hillberry. 2006. Trade Frictions and Welfare in the Gravity Model: How Much of the Iceberg Melt? *Canadian Journal of Economics* 39: 247–265.

Bergstrand, J.H. 1985. The Gravity Equation in International Trade: Some Microeconomic Foundations and Empirical Evidence. *Review of Economics and Statistics* 67 (3): 474–481.

———. 1989. The Generalized Gravity Equation, Monopolistic Competition, and the Factor–Proportions Theory in International Trade. *Review of Economics and Statistics* 71 (1): 143–153.

Blyde, J.S. 2000. Does international trade hurt the environment? Old theory, new developments. *International Trade Journal* 14 (4): 343–353.

Cheng, I.H., and H.J. Wall. 2005. Controlling for Heterogeneity in Gravity Models of Trade and Integration. *Federal Reserve Bank of St. Louis Review* 87 (1): 49–63.

Coondoo, D., and S. Dinda. 2002. Causality between income and emission: A country group-specific econometric analysis. *Ecological Economics* 40 (3): 351–367.

Copeland, B.R., and M. Scott Taylor. 2004. *Trade and Environment*. Oxford: Oxford University Press.

Dean, Judith M., Mary E. Lovely, and Hua Wang. 2009. Are Foreign Investors Attracted to Weak Environmental Regulations? Evidence from China. *Journal of Development Economics* 90 (1): 1–13.

Deardorff, Alan V. 1984. Testing Trade Theories and Predicting Trade Flows. In *Chapter 10, Handbook of International Economics*, ed. R.W. Jones and P.B. Kenen, vol. 1, 1st ed., 467–517.

———. 1995. Determinants of Bilateral Trade: Does Gravity Work in a Neoclassical World?, *NBER Working Papers* No. 5377.

Dinda, S. 2004. Environmental Kuznets Curve Hypothesis: A Survey. *Ecological Economics* 49: 431–455.

———. 2011a. Trade Opportunities for Climate Smart Goods and Technologies in Asia, paper presented at MSM 1st Annual Research Conference Nov 11–12, 2011.

———. 2011b. Climate Change and Development: Trade Opportunities of Climate Smart Goods and Technologies in Asia, *MPRA Paper* No. 34883.

———. 2014a. Climate Change and Trade Opportunity in Climate Smart Goods in Asia: Application of Gravity Model. *The International Trade Journal* 28 (3): 264–280.

———. 2014b. Climate Change: An Emerging Trade Opportunity in South Asia. *South Asian Journal of Macroeconomics and Public Finance* 3 (2): 221–239.

———. 2015. Climate Change, Trade Competitiveness, and Opportunity for Climate Friendly Goods in SAARC and Asia Pacific Regions. In *Handbook of Research on Climate Change impact on Health and Environmental Sustainability*, ed. S. Dinda, 519–542. Wisconsin, USA: IGI Global Publishers Inc.

Disdier, A.C., and K. Head. 2008. The puzzling persistence of the distance effect on bilateral trade. *Review of Economics and Statistics* 90 (1): 37–48.

Drysdale, P., and R. Garnaut. 1982. Trade Intensities and the Analysis of Bilateral Trade Flows in a Many-Country World: A Survey. *Hitotsubashi Journal of Economics* 22 (2): 62–84.

Drysdale, P., and Xu, X. 2004. Taiwan's Role in the Economic Architecture of East Asia and the Pacific, *Pacific Economic Papers* No.343.

Drysdale, P., Kalirajan, K.P., Song, L., and Huang, Y. 1997. *Trade Among the APEC Economies: An Application of a stochastic varying coefficient gravity model*, Paper presented at 26th Economists Conference, University of Tasmania.

Drysdale, P., Y. Huang, and K.P. Kalirajan. 2000. China's Trade Efficiency: Measurement and Determinants. In *APEC and liberalisation of the Chinese economy*, ed. P. Drysdale, Y. Zhang, and L. Song, 259–271. Canberra: Asia Press.

Egger, P. 2002. An Econometric View on the Estimation of Gravity Models and the Calculation of Trade Potentials. *The World Economy* 25 (2): 297–312.

Eichengreen, B., and D. Irwin. 1998. The role of history in bilateral flows. In *The regionalization of the world economy*, ed. J.A. Frankel, 33–57. Chicago: University of Chicago Press.

Frankel, J.A., E. Stain, and S.J. Wei. 1997. *Regional trading blocs in the world economic system*. Washington, D.C.: Institute for International Economics.

Gaulier, Guillaume, Mayer, Thierry and Zignago, Soledad. 2004. *"Notes on CEPII's distances measures"*, www.cepii.eu.

Ghosh, S., and S. Yamarik. 2004. Are Regional Trading Arrangements Trade Creating?: An Application of Extreme Bounds Analysis. *Journal of International Economics* 63 (2): 369–395.

Grossman, G.M., and A.B. Krueger. 1991. *Environmental impacts of the North American free trade agreement*. NBER. Working paper 3914.

———. 1995. Economic Growth and the Environment. *Quarterly Journal of Economics* 110 (14): 353–377.

Harrigan, J. 2001. Specialization and the Volume of Trade: Do the Data Obey the Laws?, National Bureau of Economic Research, *NBER Working Papers,* 8675.

Helpman, E. 1987. Imperfect Competition and International Trade: Evidence from Fourteen Industrial Countries. *Journal of the Japanese and International Economies* 1: 62–81.

Helpman, Elhanang, and Paul R. Krugman. 1985. *Market Structure and Foreign Trade.* Cambridge, MA: MIT Press.

Jha, Veena. 2008. Environmental Priorities and Trade Policy for Environmental Goods: A reality Check, *ICTSD Issue Paper* No 7, September.

———. 2009. Climate Change, Trade and Production of Renewable Energy Supply Goods: The Need to Level the Playing Field, *ICTSD Paper.*

Kalirajan, K. 1999. Stochastic Varying Coefficients Gravity Model: An Application in Trade Analysis. *Journal of Applied Statistics* 26 (2): 185–193.

Kalirajan, K., and C. Findlay. 2005. *Estimating Potential Trade Using Gravity Models: A Suggested Methodology.* Tokyo: Foundation for Advanced Studies on International Development.

Khatun, F. 2010. *Trade negotiations on environmental goods and services in the LDC context.* UNDP Discussion Paper. Bureau of Development Policy.

Learner, E.E., and J. Levinson. 1995. International trade theory: The evidence. In *Handbook of international economics,* ed. G. Grossman and K. Rogoff, vol. 3, 1339–1394. North-Holland: Elsevier.

Liddle, B. 2001. Free Trade and the Environment-Development System. *Ecological Economics* 39 (1): 21–36.

Linnemann, H. 1966. *An Econometric Study of International Trade Flows.* Amsterdam: North Holland Publishing Company.

Mani, M. 2014. *Greening India's Growth: Costs, Valuations and Trade-offs.* London: Routledge.

Markusen, J.R. 2013. Putting per-capita income back into trade theory. *Journal of International Economics* 90: 255–265.

Martínez-Zarzoso, I., and F. Nowak-Lehmann. 2003. Augmented gravity model: An empirical application to Mercosur– European trade flows. *Journal of Applied Economics* 6 (002): 291–316.

McCallum, J. 1995. National Borders Matter: Canada–U.S. Regional Trade Patterns. *American Economic Review* 85 (3): 615–623.

Meyer-Ohlendorf, Nils and Gerstetter, Christiane. 2009. Trade and Climate Change – Triggers or Barriers for Climate Friendly Technology Transfer and Development? Friedrich-Ebert-Stiftung, Occasional Paper No 41/February 2009, Berlin.

Monkelbaan, J. 2011. *Trade Preferences for Environmentally Friendly Goods and services.* Geneva: ICTSD Global Platform on Climate Change, Trade and Sustainable Energy, International Centre for Trade and Sustainable Development.

Mukhopadhyay, K., and D. Chakraborty. 2005. Is Liberalization of Trade Good for the Environment? Evidence from India. *Asia-Pacific Development Journal* 12 (1): 109–136.

Nguyen, Van Son, and K. Kalirajan. 2015. Export of Environmental Goods: India's Potential and Constraints. *Environment and Development Economics* 21: 158–179.

Nilsson, L. 2000. Trade Integration and the EU Economic Membership Criteria. *European Journal of Political Economy* 16: 807–827.

OECD/Eurostat. 1999. *The Environmental Goods and Services Industry: Manual on Data Collection and Analysis.* Paris: OECD.

Porter, M.E., and Claas van der Linde. 1995. Toward a new conception of the environment-competitiveness relationship. *The Journal of Economic Perspectives* 9 (4): 97–118.

Rauch, J. 1999. Networks versus markets in international trade. *Journal of International Economics* 48 (1): 7–35.

Ravallion, M. 2003. On Measuring Aggregate "Social Efficiency", *World Bank Policy Research Working Paper* No. 3166.

Ravenstein, E.G. 1889. The laws of migration. *Journal of the Royal Statistical Society* 52 (2): 241–305.

Rose, R. 2000. Getting things done in antimodern society: Social capital networks in Russia. In *Social Capital: A multifaced perspective*, ed. P. Dasgupta and I. Seragilden. Washington, DC: World Bank.

Rose, A.K. 2004. Do We Really Know That the WTO Increases Trade? *American Economic Review* 94 (1): 98–114.

———. 2005. Which International Institutions Promote International Trade? *Review of International Economics* 13 (4): 682–698.

Selden, T., and D. Song. 1994. Environmental quality and development: is there a Kuznets Curve for air pollution emissions? *Journal of Environmental Economics and Management* 147: 147–1614.

Stern, D.I. 2004. The rise and fall of the Environmental Kuznets Curve. *World Development* 32 (8): 1419–1439.

Tinbergen, J. 1962. *Shaping the World Economy: Suggestions for an International Economic Policy*. New York: The Twentieth Century Fund.

UNESCAP. 2007. Trade Statistics in Policy Making: A Handbook of Commonly used Trade indices and Indicator, Prepared by Mia Mikic and John Gilbert, Bangkok. Bangkok/New York: Oxford University Press.

———. 2010. *Asia pacific trade and investment report 2010: Recent trends and developments*. Bangkok/New York: Oxford University Press.

———. 2011. *Asia pacific trade and investment report 2011: Post crisis trade and investment opportunities*. Bangkok/New York: Oxford University Press.

United Nations. 2003. *UN special report on mainstreaming adaptation to climate change in Least Developed Countries (LDCs)*. Nottingham: Russell Press.

World Bank. 1992. *World Development Report*. New York: Oxford University Press.

———. 1994. *World Development Report*. New York: Oxford University Press.

———. 2008. *International Trade and Climate Change: Economic, Legal and Institutional Perspectives, the World Bank*. New York: Oxford University Press.

Index